William Ford Stanley

Notes on the Nebular Theory

William Ford Stanley
Notes on the Nebular Theory
ISBN/EAN: 9783337395889
Printed in Europe, USA, Canada, Australia, Japan
Cover: Foto ©berggeist007 / pixelio.de

More available books at **www.hansebooks.com**

NOTES

ON THE

NEBULAR THEORY

IN RELATION TO

STELLAR, SOLAR, PLANETARY, COMETARY,

AND GEOLOGICAL PHENOMENA.

BY

WILLIAM FORD STANLEY,

F.R.A.S., F.G.S., F.R.Met.Soc., M.Phys.Soc.,

AUTHOR OF TREATISES ON THE "PROPERTIES AND MOTION OF FLUIDS,"
"SURVEYING INSTRUMENTS," "DRAWING INSTRUMENTS," ETC.

"A cause qu'il ne convient pas si bien à la souveraine perfection qui est en Dieu de le faire auteur de la confusion que de l'ordre, et aussi que la notion que nous en avons est moins distincte, j'ay crû devoir icy préférer la proportion et l'ordre à la confusion du Chaos."—*Descartes.*

LONDON:
KEGAN PAUL, TRENCH, TRÜBNER & CO., Ltd.
1895.

PREFACE.

THE subject of the following pages as a study commenced to fascinate me in my youth, when, deeply imbued with the study of Newton, I was trying vainly to unravel the theoretical possibilities of the original structure of the marvellous mechanism of the universe. It has been ever present with me throughout a busy life and has been often reconsidered, leading me to the conclusion that a modified form of the Nebular Theory of Laplace might be established on some new ideas which I formed and by certain calculations that I felt sure the actual conditions warranted. These speculations are now offered to the scientific world for approval.

I arranged this matter as it presented itself to my mind originally in papers upon separate parts of the subject, as a less confident mode of introduction; but I was advised by orthodox authorities that such papers were too speculative to communicate to the learned societies that I thought at the time best adapted for their consideration. These papers, which I now edit in abstract, have been put aside for many years. Upon the permissible borders of the subject I read a paper before the Geologists' Association in March 1883, "Upon the Causes of the Elevation and Depression of the Earth's Surface," which I considered to represent the continuity of effects of nebular conditions upon the Earth (reported in 'Nature,' March 29th, 1883). Some of the speculative unpublished matter of this paper is included in the following pages, with copies of my diagrams.

I wrote a paper in 1878 upon "Some Hypothetical Conditions of the Properties and Motions of Comets," that I assume to depend upon original nebular conditions, which, after some correspondence with a high authority, I sent to the 'English Mechanic,' a journal in which astronomical subjects are frequently discussed (published 22nd June, 1883). This matter is incorporated herein with some conclusions arrived at after further consideration of the subject.

At the British Association in 1883 I read a paper entitled "Notes upon the Rotation-period of the Earth and Revolution-period of the Moon, deduced from the Nebular Hypothesis of Laplace" (Brit. Assoc. Reports, 1885, p. 915). The subject of this paper is more fully treated in the present work.

I also read a paper before the Geological Society showing the error in Mallet's theory of the contraction of the Earth in cooling, which process was assumed to produce the great elevation and inclination of strata observed in nature (Phil. Trans. vol. clxiii. part 1). This theory, which has been very generally accepted, derived most of its support, I think, from a large accidental error which I was able to point out in the mathematical calculations (Proc. Geol. Soc. June 1884), so that the causes of elevation and inclination of strata discussed herein as a process of the continuity of nebular conditions may be said to remain unexplained by any hypothesis founded upon correct data.

I have also read several papers before the British Association and the Geological Society contravening some points in the popular theory of a universal glacial age, which I think was only local at any period, and is opposed to the Nebular Theory, which I thought, at the time I wrote the papers, demanded a uniform decrement of heat in time in the entire cosmic system. This idea as regards the periodic amount of solar radiation to the Earth I have somewhat modified in these pages by considering the effects of critical temperatures upon the solar nebula; but as I have endeavoured to bring the whole subject together as briefly as possible as it presents itself to my mind, it is unnecessary to discuss more particularly what I have already attempted to do in this direction. The greater part of this matter has remained unpublished, except partly in short abstracts, being at present somewhat out of concord with prevailing theories.

What I most regret in this matter is that I am unable to fully discuss many theoretical ideas that have been proposed by scientific men without extending this sketch very much beyond the limit necessary for the brief statement of my own ideas. This is unfortunate, as I find in reading up the subject, for the most part since I wrote these pages, that other ideas approach my own in a few particulars.

One feels also that, in the discussion of a speculative subject, in proportion as ideas are original they must be difficult to correlate with the more or less established scientific theories. The whole subject, however, is undoubtedly in a tentative state, and must be studied generally upon a broader and more exact basis in detail than it has heretofore been, if a satisfactory theory is to be established. For this investigation some acceptable data must be found before refined mathematical analysis can be of much value. For this theory I can only hope I have put forward some available suggestions.

I am indebted to Mr. W. T. Lynn, B.A., F.R.A.S., for critical examination of my proofs both for reading and for calculations. I am indebted to the kindness of Mr. William Crookes, F.R.S., and Professor G. F. Fitzgerald, F.R.S., for some suggestions which have enabled me to render the matter in the second and twelfth chapters more definite and logical. I am indebted to an eminent practical geologist, who withholds his name, for reading and suggestion in

Chapters X. to XVI.; to Mr. Charles Kirk for care in the reproduction of the Plates; and to Mr. W. Francis, of my printers' firm, for care in correcting my proofs.

South Norwood, March 1895.

CONTENTS-INDEX.

CHAPTER I.

INTRODUCTION.—HISTORICAL NOTES :—Ancient Ideas—Renaissance, p. 2. Descartes—Wright, p. 3. Kant, p. 4. Lambert—Sir Wm. Herschel, p. 6. Laplace, p. 7. Mayer, p. 9. Helmholtz, p. 10. Lane—Faye, p. 11. Projectile Theory—Limits of Theory to be discussed, p. 15.

CHAPTER II.

CONDITIONS OF MATTER POSSIBLY ACTIVE IN THE CONDENSATION OF AN EXTENSIVE SYSTEM OF COSMIC NEBULÆ :—Time and Space, p. 17. Original State of Matter, p. 18. Separation of Systems of Original Matter, p. 19. Comparison with Astronomical Nebulæ, p. 20. Tenuity of Original Matter, p. 21. Milky Way—Atomic Theory, p. 22. Sir Benj. Brodie's Theory of Original Matter, p. 23. Suggestions for the Constitution of the Nebulæ—

Pneuma, p. 23. Lockyer's Dissociation — Crookes's Fractionation, p. 25. Pneumites, p. 26. Chemical Action in a Pneuma System, p. 30. Cohesion of Matter, p. 32. Observation of Astronomical Nebulæ, p. 35. Condition of the Sun, p. 36.

CHAPTER III.

FORMATION OF STELLAR AND SOLAR SYSTEMS :—Formation of Nebulæ from the original Pneuma System, p. 38. Suggested Motive Conditions in the original Pneuma, p. 40. Cyclone-inducing Conditions, p. 42. Formation of Spiral Nebular Systems, p. 45. Solar-Planetary inducing conditions in Nebulæ, p. 46. Permanency of Stellar Systems, p. 47. Distances under which Gravitation may be active—Action of Gravity in formation of Circular Orbits, p. 49.

CHAPTER IV.

STELLAR AND SOLAR CONDENSATION.—FORMATION OF ORBITS: —Separate Stellar and Solar-Planetary inducing conditions according to amount of Rotation, p. 52. Limits of a Solar-Planetary-Cometary System, p. 53. Action of Gravity on distant Condensations, p. 55. Condensation to a Solar Centre, p. 56. Direction of Approach to the Sun of exterior Matter—Formation of Orbits, p. 58. Cometary Orbits, p. 60. Formation of a Planetary Plane, p. 62. Planets formed at the Perihelion of Cometary Orbits, p. 63.

CHAPTER V.

DISCUSSION OF THE MECHANICAL PRINCIPLES UPON WHICH OUR SOLAR-PLANETARY SYSTEM MAY HAVE BEEN FORMED.—DEMONSTRATION OF THE THEORY OF LAPLACE, WITH SOME MODIFICATIONS.—LIMITS OF A COMETARY SYSTEM:—Energy of the Solar System—Its Asymmetry, p. 65. Scheme of a Symmetrical Gaseous Solar-Planetary System, p. 67. Modes of Condensation of Interior Planets in a Spheroidal Nebular System, p. 71. Mode of Condensation of Extreme Outer Solar Nebula, p. 72. Breaking-up of Gaseous Zone-Systems, p. 73. Influencing Conditions of Planet-formation—Critical Temperatures, p. 74. Some Modifying Conditions, p. 77.

CHAPTER VI.

CERTAIN CONDITIONS IN THE EARLY SOLAR SYSTEM WHICH MAY BE INFERRED FROM THE DISTANCES AND MASSES OF THE PLANETS UPON THE NEBULAR THEORY :—The Distances of the Planets from the Sun—Bode's Law, p. 80. Masses of the Planets, p. 81. Proportional Densities, p. 82. Probable Form of the Original Planetary Nebula, p. 84. Effects of the voluminous Ring of Jupiter upon the Intra-Jupiter Solar System, p. 85. Relative Rate of Cooling of the Intra-Jupiter System—with the Earth, p. 86.

CHAPTER VII.

SUGGESTIONS FOR CAUSES OF DIRECTION OF ROTATION OF THE SUN AND PLANETS.—THE DIRECTION OF REVOLUTION.—VELOCITY OF ROTATION OF THE PLANETS AND OF THEIR SATELLITES :—Direction of the Solar Axis of Rotation, p. 88. Direction of the Planets' Axes, p. 89. Rotation of the Sun, p. 90. Momentum of a Planet, p. 95. Orbital Velocity of a Planet, p. 96. Rotation of Planets, p. 97. Calculated Rotation of Jupiter and Saturn, p. 101. State of Jupiter and Saturn, p. 103. Rotation of the Asteroids—of Mars, p. 106. Rotation of the Earth, p. 108.

CHAPTER VIII.

REVOLUTION OF SATELLITES.—DIRECT MOTION.—RETROGRADE MOTION.—COMPARISON OF THE REVOLUTION OF THE MOON WITH THE ROTATION OF THE EARTH :—Revolution of Satellites with Direct Motion, p. 110. Calculation of Revolution of the Satellites of Jupiter and Saturn, p. 111. Satellites of Mars, p. 113. The Moon, p. 114. Retrograde Motion of Satellites, p. 116.

CHAPTER IX.

COMETS CONSIDERED AS ORDINARY GRAVITATIVE MATTER IN ROTATION CONSTRUCTIVELY AS A PART OF THE PLANETARY SYSTEM :—Discussion of Principles, p. 121. Comets of Long Period, p. 125. Comets of Short Period—Symmetrical Elements of Comet-formation, p. 126. Comets considered as Gravitative Matter, p. 128. Conditions under which a Comet may be considered as a Planetary Body, p. 130. Heat—Electricity—Orbital Momentum, p. 131. Elongation of the Cometary Mass near Perihelion, p. 136. Orbits of the outer parts of a Comet—Focal Point, p. 139. Direction of a Cometary Train in relation to the Sun, p. 140. Widening Curvature of the Train by crossing Orbits, 142. Formation of a New Head to a Comet, p. 144. Some general conditions, p. 145.

CHAPTER X.

THE EARTH, CONSIDERED IN EVIDENCE OF FORMER NEBULAR CONDITIONS. — ITS INTERNAL FLUIDITY.—TIDAL FRICTION.—FIGURE DUE TO ROTATION :—The Earth considered as a Model Planet, p. 147. Factors of Earth-formation, p. 148. From a Gaseous System, p. 149. From aggregation of Planetoids, p. 150. Internal Fluidity of the Earth discussed, p. 150. Effects of Tidal Friction, p. 156. Change of figure due to Rotation-Velocity, p. 157.

CHAPTER XI.

SUPERFICIAL CONDITIONS OF THE EARTH DUE TO DISCRETE CONDENSATIONS, FORMED BETWEEN THE EARTH'S ORIGINAL NEBULA-ZONE AND THE ORBIT OF MARS :—Formation of Land-areas by Inclusion of Planetoids, p. 160. Projection of a large Intra-Mars Planetoid upon the Earth, p. 165. Formation of a Continent therefrom, p. 166.

CHAPTER XII.

Hypothesis of the Formation of the Earth under purely Nebular Conditions :—Conditions specialized, p. 169. Distinct Periods of Deposition—Period of Condensation of highly Refractory Matter, p. 170. Period of Condensation of Volatile Metals, Oxygen, and Halogens with Metals and Metalloids, p. 174. Distribution of Land-areas, p. 180. Period of Deposition of Water, p. 184.

CHAPTER XIII.

Conditions of the Cooling Earth due to Formation of Ice at the Poles :—Pre-Glacial and Glacial Periods, p. 189.—Distribution of Ice at the Poles, p. 192. Present Conditions brought about by Deposition of Ice, p. 193. Where Ice-pressures are most active, p. 197. Extrusion of Volcanic Matter through Ice-pressures, p. 199. Pressure of Water and Steam in Volcanic Eruptions, p. 201. Notes upon Theories proposed p. 207.

CHAPTER XIV.

Periodic Condition of Earth-formation produced by the Effects incidental to the Nebular Clouding at Inferior Planetary Formation and at Critical Temperatures of Matter surrounding the Sun :— Condition of the Sun during Earth-formation, p. 210. Distinct Solar-heating Periods, p. 213. Division of Special Periods, p. 214. General effects of the Large Nebulous Sun upon Meteorological Conditions, p. 218.

CHAPTER XV.

CONSIDERATION OF TIME ELEMENTS IN THE SOLAR SYSTEM, PARTICULARLY FOR ESTIMATING THE AGE OF THE EARTH AND FOR ITS GEOLOGICAL PERIODS :—Time of Condensation of the Solar System, p. 220. Calculation of Time of Condensation for the Nebula extending to the Orbit of Neptune, p. 222. The same for the Orbit of the Earth, p. 223. Distribution of Time upon the Earth, throughout the Varying Periods of Condensation of the Sun and the Inferior Planets, p. 224. Table in millions of years—Period of the Formation of the Nebulous Earth, p. 225.

CHAPTER XVI.

GEOLOGICAL PERIODS CORRELATED WITH ASTRONOMICAL PHENOMENA :—Discussion of Geological Periods, p. 228. Archæan Period, p. 231. Archæan Time in Relation to the Conditions of Animal Life, p. 233.—Cambro-Silurian Period, p. 235. Devonian Period, p. 239. Permian Period, p. 241. Triassic and Rhætic Periods—Jurassic Period—Cretaceous to Tertiary Periods, p. 243. Glacial Period, p. 246. Future Period, p. 249.

APPENDICES.—A. Hypothesis of Radiation of Heat and Light, p. 252. B. Mallet's Theory, p. 256. C. Contemporaneous Stratification of Rocks of the prevailing Chemical Elements, p. 258.

NOTES

ON THE

NEBULAR THEORY.

CHAPTER I.

INTRODUCTION.—HISTORICAL NOTES.—LIMITS OF THE THEORY TO BE DISCUSSED.

THE origin of the material universe has occupied the deepest thoughts of many of our most profound thinkers. These speculations must, however, rest for ever upon the borderland of Science, where few practical men may care to tread. To bring this subject more nearly into co-relation with practical science it is necessary that we keep more nearly to the evidences of natural phenomena for our data than has heretofore been done by our scientists, who have so often in this matter followed pure speculations only. We may also possibly with advantage make tests of the theories that have been proposed by submitting them to calculations taken directly from the premises offered in the various hypotheses. To place this matter before the reader clearly for discussion, it will be convenient to offer a few historical notes on the leading theories that have been proposed, upon which it is my purpose to graft certain suggestions and to make certain calculations. These historical outlines will save space being taken by constant definitions of the earlier theories, by making use of references to the paragraph numbers to be found in this chapter.

1. The ancient ideas of the Kosmos in no way approach the possible conditions of a Nebular theory—the extent of the universe in early times was conceived to be only that which appeared evident to the senses, the earth being taken as the centre of the universe, surrounded and enclosed by firmaments that were assumed to be revolving solid constructions, which were variously defined. To this, however, we have some exceptions. Anaximines believed stars to be of fiery substance and to carry invisible earthly bodies with them. As a preliminary idea of early nebular conditions he held that air was the original material of the universe, from which all things were engendered and into which they resolve*. Pythagoras taught his disciples that the sun was the centre about which the planets revolved †, by which he accounted for eclipses and the motions of the planets, and thereby clearly anticipated what we term the Solar System of Copernicus. This was a great advance upon the prevailing theory, which was limited to a firmament or a number of crystal spheres surrounding the earth. The theory of Pythagoras, probably derived from the Chaldeans, was, however, far too much in advance of the age to be accepted by the following generations of popular scholars and theorists, who were, in many instances, strongly prejudiced against it owing to the influence of the prevailing superstitions of the time, in accordance with which alone popularity could be attained.

2. The earliest suggestion of a nebular hypothesis that occurs within the Renaissance period is probably that of Tycho Brahe, who, to account for the new star which appeared in 1572, suggested that stars were formed by condensation of

---

* 'History of Philosophy,' T. Stanley, 4th ed. 1743, p. 54.
† *Id.* p. 444.

the ethereal substance of which he imagined the Milky Way was composed *. Kepler accepted Tycho Brahe's idea, which he somewhat extended in his account of the new star which appeared in 1604, by suggesting that the nebular substance might not be confined to the Milky Way alone but may have pervaded all space †. In an account of the eclipse of the sun at Naples in 1605 he suggests that nebular matter is of the same kind as that which appears around the dark body of the moon in a total eclipse.

3. Descartes appears to have been the first to attempt to construct a complete theory of the origin of the known universe, or to systematize the matter of space. In this he assumes that universal matter originally existed in three states—coarse, fine, and very attenuated; that it drifted originally in a complex system of whirls, and that each of these whirls formed a solar or planetary system ‡. This theory, brought forward again recently by M. Faye, will be presently discussed: we may term it the *Vortex Theory.*

4. Thos. Wright was the first to suggest a complete gravitation theory of the universe founded upon astronomical observations taken accurately enough to be of any scientific value §. He suggests that the universe, represented by the Milky Way, is a unit gravitation system in general revolution in the form of a bifurcating stratum composed of all the visible stars, which resemble our sun. This is now commonly termed the *Grindstone Theory.* He estimated the direction in which our sun is travelling among the stars by discussion of the parallax. He suggests that the stars, by uniformity of creation, have revolutionary subsidiary systems of planets similar to our own solar system. In the plate which forms

---

* 'Progymnasmata,' 1572, p. 795.
† 'Stella nova in pede Serpentarii,' 1606, p. 115.
‡ 'Essais Philosophiques,' 1637. 'Speciminæ Philosophicæ,' 1644.
§ 'An Original Theory of the Universe,' Thomas Wright, 1750. See also De Morgan, Phil. Mag. 3 ser. xxxii. p. 241.

our frontispiece he shows by shading the theoretical gravitation influences over matter of our Sun, Sirius, and Rigel, in which each of these star systems is shown extending its influence over an approximately equal area of space, and nearly meeting the system of each of the others. He suggests that with more perfect telescopes the rings of Saturn will be discovered to be formed of small satellites. He considers the measurable visible sun to consist of a vaporous and a nebulous atmosphere, the dense solid or liquid body of the sun being much smaller than it appears, possibly only of about two thirds its apparent dimensions. He contends that the Milky Way forms one vast system composed of solar systems like our own. He does not appear to know of more than the six nebulæ mentioned by Halley as "light coming from an extraordinary large space in the ether, through which a lucid medium is diffused, which shines with its own proper lustre"[*]. Wright refers to these *cloudy spots* as "condensations of vapour among the mass of stars to which our sun belongs." Comets are suggested to have elliptical closed orbits, as represented in the frontispiece, and therefore to be periodical.

5. Wright's original bold but (as regards particulars left unnoticed) somewhat indefinite outline was filled up more in detail by Kant, who fully recognizes the speculations of "Wright of Durham," and accepts his general principles concerning the structure of the universe. Kant's original speculations given in the second part of his work are principally directed to account for the formation of the solar system, the mass and motion of which are assumed to have been produced by the aggregation of free particles that were formerly uniformly distributed in space in an attenuated form. The particles falling together at an early period by initial gravitation formed masses by local condensations.

---

[*] Halley, Phil. Trans. 1714.

These masses, under universal gravitation, are assumed to have encountered the resistances of other masses and particles, generally distributed, so that those parts of the system only could continue to move freely and form concentric systems which acquired a linear velocity sufficiently in equation with the nearest centralizing attraction to produce orbital motion. The matter deflected from the direct line of attraction towards the sun passed into revolution about it. The revolution produced a denser extended equatorial plane, into which exterior matter was drawn whilst approaching the sun. The velocity of the particle falling towards the sun depended upon the distance fallen, the direction it finally took upon the sum of lateral deviations it experienced in consequence of encounters with other particles, which directed it, under the influence of gravitation, into the path of least resistance. The particles which did not meet the conditions of circular or orbital motion fell into the sun, where his attraction predominated. The particles deflected or held in equilibrium in nearly the same orbit formed the planet by gravitating towards denser condensations of surrounding matter, which, acting under similar conditions, took the same direction of revolution as the sun. The same motive principles in matter which produced the revolution of the planet around the sun also produced its own rotation and the revolution of its satellites in the same direction. The planet in regard to rotation is considered as an independent body. The rotative movement might therefore, according to the momentum of the mean drift of its matter, take one direction or the other, the direction of rotation being due to the unequal velocity of the particles in circulation around the sun at the time they were falling upon the new forming planet through its prevailing local attraction. Saturn is taken as a particular case for consideration, in which vaporous conditions of condensation are suggested, the ring being in revolution and thrown off the planet where the centrifugal force of its matter becomes in equilibrium with gravitation *. It will be convenient to denominate the nebular theory of Kant the *Discrete System*, in contradistinction to the *Concrete* or gaseous system of Sir William Herschel and Laplace, notices of which follow.

It is readily seen that matter equally distributed in space could not possibly drift in the manner proposed by Kant, but that it must at an early period fall into the whirl system of motion proposed by Descartes. M. Faye, although generally supporting the discrete theory of Kant, has demonstrated that if matter drifted under the influence of gravitation only, as proposed by this philosopher, it would possess no rotation upon condensation in forming the sun or a separate planetary system †.

Lambert followed closely in the theory of Kant, his greatest divergence being in the division of the universe into many galactic systems of which our Milky Way represents one only ‡. This is now denominated the *Island Theory*.

6. No further advance was made in the nebular theory until over 2000 nebulæ had been discovered and examined by Sir William Herschel, an account of which was placed before the Royal Society in several papers from 1784 onwards. The nebulæ were recognized individually as immense gravitation systems in 1789. The planetary nebulæ were adduced as giving evidences of atmospheres of shining fluid about stellar foci, which were suggested to be in a state of condensation in 1791. The entire subject is brought together, embracing ideas of the origin of our own solar system being derived from nebular matter, in Phil. Trans. 1811, p. 269 *et seq.* The conclusions arrived at are somewhat less original than Herschel supposed. They possibly mark in one respect the influence of chemical discovery, in which the smallest parts of bodies were beginning to be recognized as distinctly

---

* 'Allgemeine Naturgeschichte und Theorie des Himmels,' 1755.
† 'Sur l'Origine du Monde,' 2nd ed. p. 135.
‡ 'Cosmologische Briefe,' 1761.

structural units *, all of which might be brought to a gaseous state by heat : therefore the possibility of the existence of all known matter in three forms—gaseous, liquid, and solid—dependent upon the special temperature to which any element is exposed. The recognition of a possible gaseous state for all matter appears to have suppressed for the time the former prevailing idea, as regards the nebular hypothesis, that material particles separately distributed in space represented the most attenuated form of matter. The gaseous element offered also at the same time a new foundation for the construction of a nebular hypothesis.

7. The subject is taken up by the powerful analytical mind of Laplace in 1796 † and in 1799, and advanced in following years ‡. It is treated entirely *de novo*, this author evidently not knowing how much had been thought out in the same direction by others. He follows without knowing it closely upon the general arguments of Kant, particularly those of his theory of the formation of the rings of Saturn, with the important addition of the introduction of the gaseous nebular element as a universal medium. He formulates that whatever could have directed the movements of the planets, it must have been originally a concrete system embracing the whole of these bodies, which could not possibly, seeing its immense extent, have been other than an aerial fluid surrounding the sun and possessing the same direction of revolution. To ensure this possible extension of matter he supposes that the nebulous gaseous matter was of sufficiently high temperature for all solids to exist in a purely gaseous state. He suggests that this attenuated matter of the solar system probably resembled some of the nebulæ visible in the telescope, or more particularly the nebulous stars which were formed from the general more attenuated and highly heated

* Lavoisier's 'Traité élémentaire de Chemie,' 1789.
† 'Exposition du Système du Monde,' vol. ii. p. 295.
‡ *Id.* 3rd edit. 1813

nebulous matter. The original momentum of the solar nebula in revolution was conserved, so that as it contracted it attained higher velocity. The planets were formed at the limits of the solar atmosphere when it was a planetary nebula by successive zones of vapour being abandoned in the plane of the sun's nebular equator, at a radius where the centrifugal force of the zone, due to its contraction and accelerated rotation, was in equation with gravity for the orbit of the planet. After separation the parts of the zone-ring would maintain the same angular velocity as they had while in contact with the sun for a time; but in falling towards the new planet forming by local condensation, the exterior matter, besides its excess of linear velocity over interior parts due to its exterior position, would attain further excess of velocity through gravity. The downward impulse of gravitation into tangential velocity would impress an excess of velocity over the original angular velocity in condensation upon the planet, and thereby cause its rotation to be in the same direction as the revolution of the planet around the sun. This effect will be discussed, with a diagram, in the body of the work. The satellites were formed at the limits of the atmosphere of the nebulous planets at an early period in the same manner as the planets themselves were formed about the sun. He suggests that comets are of another system with linked orbits, and that they move under the mutual attractions of our sun and other separate stars. The theory of Laplace has many able supporters, among the most able of whom is the celebrated astronomer M. C. Wolf[*], who has made important additions to it.

8. The discovery of certain factors of the mechanical theory of heat by Mayer in 1842 led this philosopher in 1848 to propound a theory of the possibility of the sun's heat being maintained by the percussion of the fall of meteoric

---

[*] 'Les Hypothèses Cosmogoniques,' 1886, p. 35.

matter upon his surface *, a principle entailing the formation of the sun from such matter, or what became the "*Meteoric theory of the Sun.*" This theory, which resembles that of Buffon, wherein the sun's heat is suggested to be maintained by the fall of comets constantly upon his surface, was extended to the formation of the planets also. It was thought to be supported by the resolution of many presupposed gaseous nebulæ into stars by means of the great reflecting telescope of Lord Rosse, erected in 1845. These observations led many astronomers to hold that all nebulæ would be resolvable if sufficient optical power were applied to them, and, therefore, that we possess no certain inference of the presence of incandescent gaseous matter in space.

9. Mayer's theory was fully investigated by Lord Kelvin, and its entire insufficiency to account for the dispensation of solar heat clearly demonstrated †. It was supported by Prof. Tait, who extended the conditions by suggesting that the incandescent state of nebulæ which were afterwards spectroscopically demonstrated as being gaseous matter by Dr. Huggins, might be derived from collisions of small cosmical bodies which were surrounded by gaseous atmospheres. Lord Kelvin, in reconsidering the meteoric theory with this modification, thought that, although the sun might have been formed by the cohesion of small masses, his heat could not be maintained by the mechanical collisions of gravitating matter only. He considers the low rate of cooling, and the consequent constancy of emission of heat, as being covered in great part by the high specific heat of the matter of the sun. He shows that if the earth fell directly to the sun, this would only maintain its present heat for another 95 years ‡. He states that if the whole mass of the planets were to fall into the sun from their orbits, the

* 'Dynamik des Himmels,' 1842, p. 12.
† Trans. R. S. Edinburgh, vol. xxi. p. 66.
‡ Thomson and Tait, 'Good Words,' October 1862.

heat engendered by their united collisions would only cover the emission of heat from the sun at its present rate for 45,589 years. The meteoric theory is supported by Prof. Lockyer * and Prof. A. Winchell †, and enlarged to the extent of solar heat being produced by the collision of our sun with another star by the late Dr. Croll ‡, an idea originally suggested by Sir William Herschel before his speculations upon the nebular theory §.

10. The late illustrious Helmholtz, in a lecture at Königsberg, Feb. 7, 1854, accepted the theory of Laplace, stipulating a special form of gaseous matter represented by a state of infinite diffusion in which the gas was affected by forces of mutual and central gravitation only, but was not necessarily in a heated state. Helmholtz determined the amount of heat that would be generated by this form of gaseous condensation in the sun and planets of our system up to the present time upon his theory. He states " that if we assume about the 454th part of the mechanical force remains as such, the remainder converted into heat would be sufficient to raise a mass of water equal to the mass of sun and planets 28 million degrees Centigrade." He assumes that the greater part of the heat was dissipated in space ages ago. He states that the cooling of the earth alone from a temperature of 2000 to 200 degrees Centigrade would, according to the experiments of Bischof upon basalt, require 350 million years. To convert the same matter from a nebular state would take a period beyond his conjecture. He supposes the condensation of the sun to continue by his attraction causing the falling of the surface towards the centre, and thereby producing a continual development of heat through pressure which, assuming the sun to be reduced by this

---

* 'The Meteoritic Hypothesis,' 1890.
† 'World Life,' 1889.
‡ 'Stellar Evolution,' 1889.
§ Phil. Trans. 1785, p. 213.

constant contraction to the density of the earth, would at the present rate of emission take a period of about 17 million years *, so that the sun in relation to the period of stellar life is near its point of extinction.

11. As an important factor of the effects of conservation of energy in the condensation of a nebula, a law of cooling of masses of gas was discovered by Mr. J. Homer Lane, of Washington, which is given in a paper "On the Theoretical Temperature of the Sun" †. This law is shown in the following manner :—If a globular gaseous mass is condensed to one-half its primitive diameter, the central attraction upon any part of its mass will be increased four-fold, while the surface will be reduced one-fourth. Hence the pressure per unit of surface will be increased sixteen times. Therefore, if the elastic gravitating forces were in equilibrium in the primitive condition of the gaseous mass, the temperature must be doubled in order that they may still be in equilibrium when the diameter is reduced one-half. Under these conditions the intensity of the heat of the sun must have increased with its contraction from the nebular condition. This is a most important consideration in showing the possibility of the conservation of the energy of the solar system in past time, during the continuous emission of heat from its former more extensive surface.

12. Recently the celebrated French astronomer M. Faye has written a learned work bringing forward much antique lore upon the subject ‡. He adopts the theory of discrete matter being originally dispersed in space, following the theory of Kant, and of its condensation upon the thermodynamic principles proposed by Mayer, by which the collisions of gravitating matter about the sun produce its heat and mass under certain conditions. He supposes that matter drifted

---

* Phil. Mag. ser. 4, vol. xi. p. 505 *et seq.*
† American Journal of Science, July 1870.
‡ 'Sur l'Origine du Monde,' 2nd ed. 1880.

originally in cyclones, according to the whirlpool theory of Descartes. He objects to the purely mechanical demonstrations of Laplace, whom science generally regards as one of the greatest celestial mechanics since Newton, in his theory showing the direction of rotation the planets and satellites must necessarily take upon exterior condensation through contraction of a rotating gaseous system. It must, however, be noticed in this argument that M. Faye changes the theoretical premises from a concrete or fluid system to a discrete or chaotic system wherein the particles of matter are assumed to be originally moving in free orbits *, in which it is certain that the application of the arguments of Laplace cannot hold. It does not appear to me that there is really very much difference between certain of the conceptions of Descartes which are applicable to the subject, if these are stripped of their complications, and those of Laplace. Both these philosophers negative a chaotic state †, and recognize the necessity for assuming an original fluid state to account for the direction of the rotation of the planets being the same as that of the revolution in their orbits. This gaseous state, surrounding the solar system in which grosser particles are assumed to float, is defined by Descartes as " ciel liquide dont les parties sont extrêmement agités " ‡, also as " corps subtile et très liquid " §,—conceptions of a gas which do not vary greatly from those of Clausius and Clerk-Maxwell. With Laplace the original solar nebula moved in all parts with equal angular velocity; with Descartes its motion was cyclonic, which is the only possible form of motion for a fluid system moving with uniform angular rotation, while condensing or contracting under gravitation, to enable it to find accommodation for the momentum of its separate parts,

* ' Sur l'Origine du Monde,' 2nd ed. 1880, p. 264.
† ' Les Principes de la Philosophie,' p. 147.
‡ *Id.* p. 129.
§ *Id.* p. 184.

as I have shown in principle in my work on the Motion of Fluids *.

13. One most important work that M. Faye has done in this theory is to show that a discrete system of matter distributed in space, in orbital motion in each of its separate parts, will, upon condensation under the action of gravitation to form an exterior body or planet, cause this body to rotate in the reverse direction to that of its solar orbital motion †. This demonstration removes a difficulty formerly experienced in the acceptance of the theory of Laplace since the discovery of the reverse direction of revolution of the satellites of Uranus and Neptune. It further shows the probability that the widely attenuated nebulous matter which may have been present in space as a part of our solar nebula exterior to the orbit of Saturn may have condensed at an early stage into free particles before the concrete formation of the planets Uranus and Neptune and their satellites. This may be taken as a very probable hypothesis, the possibility of which appears to have escaped the powerfully analytical mind of Laplace.

14. In the hypothetical element of our knowledge of the extent of time which may have been taken in solar-planetary formation, M. Faye leaves the astronomical to consider the geological conditions by taking the earth's superficial stratification as an index of the entire past of the solar system. This is unfortunate for one who has evidently not made geology a serious study. In this discussion past time is divided into Eocene, Miocene, and Pliocene periods, all of which the geological student regards as recent ‡. This error, in a geological sense, is discovered and corrected in a second edition of his important work, but in this case it still

---

* 'Experimental Researches into the Properties and the Motions of Fluids' (Spon, 1881), p. 224 *et seq.*

† 'Sur l'Origine du Monde,' 2nd ed. 1885, p. 117.

‡ 'Sur l'Origine du Monde,' 1st ed. p. 254.

compresses solar-planetary time into 20 million years. This period has been proposed by eminent physicists upon arbitrary data, but is accepted by very few practical geologists as sufficient. Prof. John Perry has quite recently suggested for consideration much more probable data for the time of cooling of the earth to its present state by taking it from a uniform temperature of $7000°$ Fahr., by which, upon his calculation, the time would be extended to 100 million years [*].

Geological time, if the evidences taken from fossil remains, the stratification of miles in thickness by slow deposition of rocks, the removal of these rocks by erosion many times, with erasure of faulting due to plutonic action, are fairly considered from observation and taken altogether, appears to carry the limit of time beyond possible conception. No short period of a few million years will satisfy the evidences of the changes occurring in the evolution of animal life alone—the changes from one stratum to another presenting gaps evidently of much longer period than the period of deposition. If we take the earliest life we know of, the animal is still a perfect highly organized structure, which for possible evolution indicates a long period lost to observation in fossil remains. No one can estimate the evidences of geological time unless he works in the field of geology. Let him follow in the excursion of such excellent societies as that of the Geological Association of London, stand in front of a mountain of an early Silurian period, formed in part entirely of comminuted shells of mollusks, with here and there a perfect weathered specimen. He begins to feel the immensity of time the generation of this life-refuse must have taken, although he knows his observation extends over but a small fraction of a long series of periods. The limit of infinite time is taken by some modern philosophers in the same spirit of

---

[*] 'Nature,' Jan. 3, 1895, p. 224.

doubt as the limit of space was formerly taken by the ancients.

15. No attempt will be made in these pages to discuss the hypothesis of the diffusion of matter into space by violent explosions from the sun and the earth and from other cosmic bodies, as probably this hypothesis will have but a limited time popularity. The difficulty to the serious physicist, as shown by Lord Kelvin, is to comprehend the wonderful conservation of energy we find in the sun and stars for the dispensation of heat and light. It is therefore physically beyond comprehension that there could remain in cosmic bodies the large amount of energy sufficient for the dispensation of heat and light, and at the same time an equal or greater amount of energy beyond that which is in any way evident, for the diffusion of matter such as satellites, comets, and meteorites into space by projection from their orbit foci, even if this would really account for their present motion and condition. Otherwise the projectile hypothesis is in every way in direct opposition to the nebular hypothesis which it will be my object to consider. We cannot possibly form a theory in which we derive energy from condensation, dispense it equally by diffusion and still have it largely conserved, as we know it to be actually in solar and stellar systems.

16. The theory herein treated will be directed to show the possibilities of the concentration of sufficient solar energy, and of sufficient geological time in the past, to satisfy the direct inferences of observation. This may be possibly best secured by assuming our original solar nebula to be represented by such actual nebulæ as we may observe with the telescope. The nebulæ selected will be assumed to go through certain changes upon condensation, the state of which may also be represented by other visible nebulæ. In this study we may follow closely in the theory of Laplace with development as far as possible by calculation of the actual motions of some part of the solar-planetary system.

The proposed data, founded by inferences drawn from astronomical observation, will entirely fall in with the possibility of heat and motive energy being due to concentration of gaseous matter in a state of infinite diffusion in revolution in the solar system, as proposed in the contraction theory of Helmholtz. At the same time we are bound to consider the effects of gravity acting upon discrete matter in cosmic formations, found in the theories of Kant and Faye, although such matter may have been originally formed by condensation of gaseous matter. The evidence of discrete cosmic matter rests upon the observation of the fall of meteorites possessed of planetary velocities to the earth, which is still taking place. Therefore, upon these premises there is the probability that throughout the formation of our solar-planetary system there were both gaseous and discrete condensations upon our sun and planets. This may be particularly evident in periods, in an early discrete system of condensation of the very attenuated external nebula before its planetary condensation, and in the possibility also of discrete condensations occurring at a period when the solar nebula by radiation of its initial heat fell below a temperature sufficient to support the gaseous state outside the present sun. These principles will be developed in the following pages, and the mode of special conditions of early condensation be discussed for contemporary astronomical and for geological conditions in the past. Some suggestions will be carried forward, so far as the conditions remain active, to the present, and very hypothetically only for future periods in relation to the sun and the earth.

CHAPTER II.

CONDITIONS OF MATTER POSSIBLY ACTIVE IN THE CONDENSATION OF AN EXTENSIVE SYSTEM OF COSMIC NEBULÆ.

17. *Time and Space.*—The infinities of time and space are not definable by any mental concept. The mind can only grasp the idea of a distant period of time and of a limited space. If we could imagine for infinite space any perfectly enclosed isolated space, to be filled with incandescent nebulous or gaseous matter, such a space would continue in the same state for infinite time, as there could be no loss of heat therefrom to condense the gas by exterior radiation to form concrete liquid or solid matter, or to concentrate locally a part of the energy of the system in any manner whatever. If we please to imagine that matter originally existed in discrete particles equally distributed in infinite space under the like conditions, no centralized gravitation system could be formed. Under these conditions it becomes evident, if we adopt either the principles of the hypothesis of Laplace or that of Kant, that we require a separate volume or an original local condensation to be subject to gravitational influences to form a star or system of stars, which volume must be isolated within the vacuous space, or in the surrounding matter or ether that must be relatively of lower specific density.

18. *Proposed Gaseous State.*—It has always been considered as the groundwork of any cosmology to arrive at a clear definition of the original state of matter. If we rely upon actual observation for our data, we have evidence of

incandescent gaseous matter isolated in space in our observable nebulæ, whereas we can have no evidence of a discrete state of matter widely distributed in scattered small units of a few grains in weight, a mile or more apart, as some philosophers have suggested, as such a discrete system would be quite impossible of visual recognition. At the same time it is the probable condition that any isolated system of attenuated gaseous matter free to radiate heat into space will condense at a certain stage of temperature and form solid matter, particularly if the gas is a heated form of ordinary solid matter. Therefore, taking an original nebula to have been in a gaseous state would not, in some cases, materially change the final results from that of a discrete system as regards the probable condition of condensation which may be instituted to form the present stars, sun, or planets, if other conditions support this theory.

19. As regards the original state of matter from which we may conceive cosmic bodies were formed, it appears to be most rational for the mind to assume that matter existed originally in a pure or elementary state, and that it afterwards became mixed or combined by the action of interior and exterior forces acting upon it under principles which we generally term the laws of Nature. This purity of state for the units of attenuated matter may possibly be best conceived by taking it to be originally gaseous, as all matter can be shown experimentally to exist for indefinite time permanent in this very attenuated condition where its heat is conserved from radiation. Whereas, any system of discrete particles or dust in the presence of gravitation acting upon it cannot be kept in an attenuated or separate state without impression of an exterior force, rotative or other, to act constantly upon every particle as a means of separation. It becomes, therefore, convenient in this discussion, without regard to theory, to presume some form of a gaseous state to be the original condition, as experiment shows that all material bodies with

which we are acquainted may reasonably have been derived therefrom by reduction of temperature alone, whatever state or mass the final condensation may assume.

20. *Separation of Systems of Original Matter.*—To support the nebular theory of Sir William Herschel and Laplace, which assumes that stars and solar-planetary systems were formed from an extensive gaseous nebula widely distributed, such as we have evidence of actually in local systems of matter dispersed in the universe available to vision in the telescope, and to analysis by means of the spectroscope, it is necessary, as stated above, that we should assume a distant period of time when a certain volume or separate volumes of highly attenuated nebulous matter existed detached from surrounding space yet moving within it. The volume of such matter may be as extensive as we may please to imagine it without changing the general conditions. It may embrace the whole of that part of the universe we define as the Milky Way, or for particular evidences in detail be restricted to our own solar system. To support this theory it is only necessary that the nebula in the system that we separately define should have a surface boundary from which it can radiate the heat into space which maintained it originally in the nebulous or gaseous state. We have, by the effect of the radiation of heat from such a system of matter, the assurance of its constant contraction in volume, tending, by the approaching nearness or contiguous cohesion of its parts, to the closer accumulation of matter in a local focus or in local foci, so that matter is thereby brought more forcibly under the centralizing action of gravitation. The effects of the condensations upon such foci from a gaseous system render available, upon chemical change of state or upon thermodynamic action, a store of energy which is sufficient in the extent of nebulæ here considered for the formation of incandescent stars, and, if any limited volume of nebula be taken to be locally in rotation, for the production of a planetary system that may be finally

formed therefrom, after a certain amount of radiation of its initial heat into space.

21. When we conceive that our solar nebula may have formed a part of a general system of nebulæ as defined above, and therefore that it may have resembled other astronomical nebulæ, we may conclude that it was of that form, among the many known forms, best adapted to produce our present solar-planetary system upon condensation. Further, it is not certain that the nebula of our system, taken as a motive system, may not have gone through certain changes by which it might at various periods be represented by various forms of visible nebulæ. Thus an originally diffused system could not have a nucleus or central condensation until this was formed, and when formed the nebula would possess a new external appearance. There are known nebulæ which present irregularities of form inconsistent with condensed gravity systems, appearing as irregular streaks and masses of incandescent hydrogen and helium. Such systems we cannot assume to have arrived at their final concentrated forms, although it is at the same time probable that they are more complete as gravitative systems than they appear. Probably the marked irregularity to telescopic vision in some of these masses, particularly of the spiral nebulæ, depends upon the incandescent hydrogen and helium being the visible part of the nebulous mass, whereas other gaseous matter, invisible in an incandescent gaseous state, makes up the entire mass. Such matter for instance as hydrogen united with oxygen in the proportion to form water would be invisible under the conditions which would render the hydrogen visible. This residual matter, which the spectroscope does not grasp, may possibly be detected at a future time by some new method of analysis. Possibly it may be inferred in some cases by refraction and by clouding effects upon more distant objects. In nebulæ that are so extensive as to suggest nearness to us, refraction may be suggested as evidence of transparent

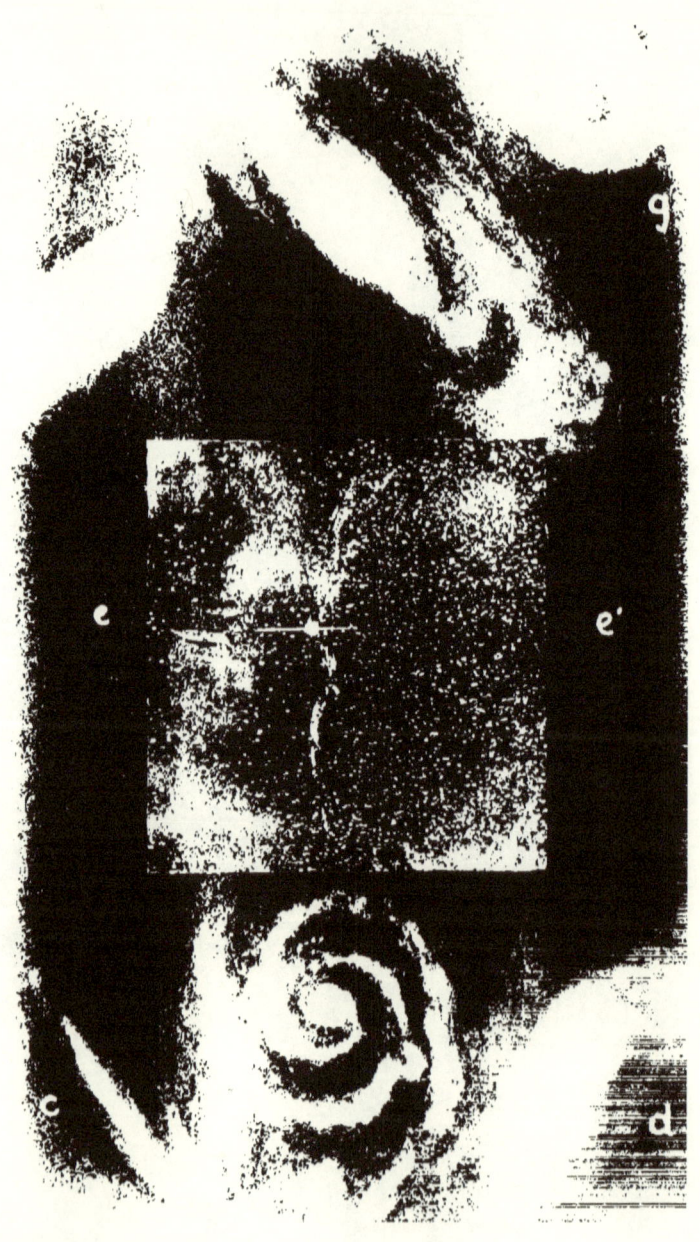

matter. In part of the nebula near 52 Cygni ♄ V 15, the visible nebula appears as a curved streak of incandescent hydrogen, shown in one of Dr. Isaac Roberts's beautiful photographs, wherein upon the hollow side of the curve stars appear to be larger and much more numerous than on the convex side, as though the complete nebula were of spheroidal or lenticular form upon the concave side, invisible itself through its transparency, yet possessing sufficient refractive power to act as the object-glass of a telescope to magnify and bring out stars in the background which would otherwise be invisible in our telescopes. Dr. Roberts states that " this gigantic nebula is of an irregular oval character ; " and that " the bright side of this nebula seems to form a sharply-defined boundary between the stream of the Milky Way stars and those on its preceding side." The photograph * for Plate II. ee' was taken by permission from a print, and does not do justice to Dr. Roberts's original negative. The hollow side where magnification occurs is shown towards e'. The dimness surrounding nebular fields being evidence of the presence of nearly invisible matter was pointed out by Sir Wm. Herschel. Possibly also such matter may surround and float up the hydrogen chromosphere of the sun.

22. *Tenuity of Original Matter.*—This is necessarily insisted upon in any system of cosmology. If we take a globular volume of gas extending to the orbit of Neptune only as the limit of our solar system and extend the mass of our sun and planets to this volume, we find by calculation that the mean density of such matter would be equal to about 1/166800000 that of air at the earth's surface. If we extend this to the mean distance between our sun and a near star, we should have to add many decimal places to our denominator. Further, it is difficult to conceive that an isolated system of matter as here proposed could remain of

* 'Selection of Photographs of Stars, Star-clusters, and Nebulæ,' p. 115.

equal density throughout, unless the central heat was enormously greater than that of the outer parts. Therefore the gaseous matter must decrease in density in some form of geometrical ratio from the centre to the exterior surface, and this must produce a tenuity about the limits much greater than that shown even by the mean density of the system suggested above.

The speculations of Sir Wm. Herschel, as well as those of Wright, infer that the entire system of the Milky Way formed at one period a unit system of matter. This appears to be probable to modern science from evidences of the unity of chemical constitution of the stars shown by the spectroscope. To account for sufficient tenuity in original cosmic matter, it was suggested by Descartes that a small mass divided into detached particles as nearly in contact as matter can approach may fill a volume however large. Infinite divisibility of matter is, however, inconsistent with chemical phenomena, which are better explained by the atomic theory.

23. In regard to the atomic theory, if we may take jointly the calculations of Cauchy from the motion of light in solids and liquids, of Lord Kelvin from certain electrical phenomena, and of Clausius and Clerk-Maxwell from gaseous phenomena,—the mean size of the ultimate atom is about one 500-millionth of an inch[*]. In the amount of diffusion discussed in the previous paragraphs for space this would leave less than a single atom to the cubic metre.

Therefore, if the above-stated gaseous theory is approximately correct, it is difficult to suggest that such an atomic system formed our original nebula, even if the nebula were sufficiently heated to produce a system of general gaseous diffusion for such atomic separation. The principles of diffusion may be materially strengthened if we can find it accordant with the inferences of science that matter

---

[*] Lord Kelvin, Proc. Roy. Inst. vol. x. pt. 2, p. 213.

may exist in much finer division than in the atomic state as theoretically defined above.

Sir Benjamin Brodie in a lecture on Ideal Chemistry, delivered before the Chemical Society in 1867, suggests a former wider division of matter than that of the atomic state in a manner very applicable to our subject. He says:—"We may conceive that in remote time and in remote space, there did exist formerly, or possibly do exist now, certain simpler forms of matter than we find on the surface of the globe— a, χ, ξ, ν, and so on. .... We may consider that in remote ages the temperature of matter was much higher than it is now, and that these other things existed then in the state of perfect gases—separate existences—uncombined .... We may then conceive that the temperature began to fall, and these things to combine with one another and to enter into new forms of existence, appropriate to the circumstances in which they were placed..... We may further consider that as the temperature went on falling, certain forms of matter became more permanent and more stable, to the exclusion of other forms. ... We may conceive of this process the lowering of the temperature going on, so that these substances, when once formed, could never be decomposed— in fact, that the resolution of these bodies into their component elements could never occur again. You would then have something of our present state of things"*.

24. *Suggestions for the Constitution of the Nebulæ.— Pneuma.*—The word "nebulæ" which is used in Astronomy to define luminous cloud-like matter, particularly incandescent hydrogen, helium, and carbon, will, upon the suggestion given above, scarcely embrace the entire early attenuated form of matter that the full consideration of the nebular theory demands, as such early nebulæ must have been formed of the constituent parts of all chemical elements,

* Quoted from William Crookes, F.R.S., Address Chemical Section Brit. Assoc. 1886, Reports, p. 559.

not only hydrogen, helium, or other gases that become visible in an incandescent state under electrical excitation. I propose the word *pneuma* to specially define this most attenuated form of gaseous matter which may have pervaded space, composed of any or all the chemical elements, and which represented the state of infinite diffusion proposed by Helmholtz. This matter would be transparent and not be visible in any form except when undergoing chemical combination or in condensation to form the visible nebula. The pneuma, in condensing to form the nebula, may be assumed to develop heat and electrical excitation, which renders the nebula condensed therefrom incandescent at the time.

25. In the condensation of a pneuma system to a nebular one, through radiation of heat, it must be the *exterior surface* alone of the nebula where chemical action can take place, and where heat and electricity are developed through the condensation, which makes the nebula become in any degree visible. A further condensation of the interior of the nebula to form a central gravitation system or sun renders also this centre visible by the heat due to the pressure of the nebula upon condensation. The light from the centre passes through the transparent part of the nebula or pneuma, which may, or may not, at the time be undergoing chemical action.

26. Taking the subject more in detail to meet possible conditions, the constitution of the pneuma which appears to me the most consistent with the undulatory theory of light, and at the same time to present evidence of sufficient tenuity to unite the material constitution of star systems, is that of perfect atomic dissociation of elementary matter to the extent that every line of light or shadow, as the case may be, made visible in the spectroscope proceeding from electrically excited highly incandescent matter represents an active factor or, if we please so to term it, a distinct kind of dissociated atom in comparison with which the chemical atom may be considered to represent a mass. This appears to be a most simple

hypothesis of the original attenuated form of matter, which defines the factors a, χ, ξ, ν of Sir Benjamin Brodie. In this construction we may ascribe to each kind of pneuma atom in a free state or when excited by heat or electricity *one vibrational period only*. It also accounts, from the certainty of the great expansion in outward volume of any known matter to produce this dissociated state, for the possibility of the matter of any single star system extending originally to the matter or pneuma of other stars.

27. The dissociation of atomic matter to the extent represented by spectral lines given above was originally proposed by Prof. Lockyer as a possible state of highly heated matter made visible by special lines in the spectra of the sun and stars, not as representing the early condition of the nebular system here proposed, for which this scientist holds quite an opposite theory*. In his theory of the dissociation of matter in the sun and stars, Prof. Lockyer endeavours to show that the dissociated atoms may possibly enter into several chemical elements represented by many lines in their spectra, so that any element may lack the matter that produces certain spectral lines. This is proposed to be particularly shown in one case by the omission of certain spectral lines in the iron group of the sun and of stellar spectra †. That the same principles may be inferred of an original form of dissociation is possibly evident in a case where there is a tendency to a community of a certain system of material associates, as in the yttrium group of metals, which have been found to be possible of separation by the refined experiments on "Fractionation" by Crookes, in which he has been able to separate yttrium in its commercial state into five or perhaps eight distinct elements giving special spectral lines ‡.

28. In suggesting a new word for the pre-nebular state of

* 'The Meteoritic Hypothesis,' 1890.
† 'Studies in Spectrum Analysis,' p. 166.
‡ British Association Reports, 1886, p. 583.

matter, it may be thought that this might be expressed by some term already in use, as for instance the *primitive fluid* of Lord Kelvin *, or by one of the numerous conceptions of the *ether* as that of Prof. O. Lodge, wherein ether is conceived to be the only matter and force forming substance †, or by *protyle*, the pre-nebular matter of Crookes ‡. Such conceptions may possibly embrace the qualities of the original materials of the universe, but they are so entirely hypothetical that they bear no relation to experience, which takes its first conception from structure. The indefinitely complex and variable motivity of a fluid assumed to produce the various kinds of known matter is more difficult of conception than that of separate structural units. Further, if such units can be correlated with natural phenomena as in chemistry or spectroscopy, they become the legitimate groundwork for the elements of a theory. With this conception the idea herein intended to be expressed by pneuma is that it is an active substance composed of units which represent, separately or in combination, all the various properties of matter. These separate distinct elements, of which there are assumed to be a much greater number than that of our acknowledged chemical elements, may amount possibly to 10,000 or more factors or varieties. Upon this proposition it is more probable that chemical elements may be split up into many more elements, than that they may hereafter be reduced in number by finding any more generally specialized constituent material or atom.

For conciseness in the following discussion, the pneuma-atom, or what we may call the Lockyer-dissociation-unit, will be termed a *pneumite*. The state and associations of such pneumites will be now considered as the groundwork of the nebular theory to be proposed.

* Enc. Brit. 9th ed. vol. iii. p. 45.
† Lecture, London Institution, Dec. 1882.
‡ Address, Brit. Assoc., Chemical Section, 1886.

29. *The pneumites* at the same temperature may be all of equal size; they must be very much smaller than the atom, probably not over 1/10,000 of its diameter. They may follow the conditions of Prout's or of the Newlands-Mendelejeff periodic law and combine consistently with producing equivalents to di-hydrogen atoms in giving units of atomic weight. They have precisely the same capacity for heat. They probably possess many uniform properties which are common to all matter besides the special individual properties of each special pneumite, the active conditions of which will now be suggested in relation to our subject.

30. The action of electrical excitation or of intense or possibly original heat is assumed, as suggested by Prof. Lockyer, to have power to set any pneumite free from cohesion to other matter where this is not subject to severe pressure, causing repulsion between pneumite and pneumite. In this case a pneumite may under excitation exist as a separate free unit, if there is no surrounding pressure or local attraction acting too forcibly upon the system of which it forms a part to cause its combination. Upon its combination with other pneumites to enter the atomic state it will develop heat or electrical excitation comparable to that which would cause its separation.

In the free state of the pneumite it is assumed to possess or attain only a single rate of vibration period for each special pneumite. The vibration period may depend partly upon the initial elasticity of the surface of the pneumite, which may be partly developed upon its surface as an expansion by heat, as we know no limit to the action of this force, or as a repulsion under like sign of electricity, causing by this expansion or repulsion from contact the separation from other near pneumites to render it motive. This may also produce vibrational effects from projection through expansion causing the collision of one pneumite with others; the elastic motivity of which may continue in its vibrational effects by reflection in collisions through the reaction of its weight or energy in attraction of gravitation or affinity towards its own and other matter as qualities of the special pneumite. Any of these conditions may distinguish a pneumite as a separate special factor of matter.

31. A physical constitution of the pneumite may be suggested similar to that I originally proposed for the atom, that of a perfectly hard centre and a perfectly tough and infinitely elastic impressible and compressible coating *—an outward condition of matter that may possibly be inferred

Fig. 2.

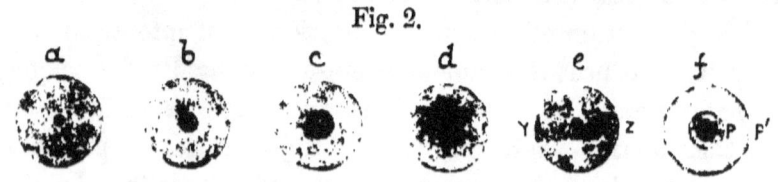

from the force required to bring two convex surfaces of glass nearly together. The pneumites of equal size and at the same temperature may possess relatively different diameters of centres and of elastic coatings, or they may possess diminishing density from the centre or polarity. Thus fig. 2 a may represent diagrammatically a very light elastic pneumite of wide or slow vibrational period, c one of rapid period, b one of intermediate period.

32. As regards the elastic sensitiveness of any pneumite, d may represent a very sensitive form in which the coating diminishes in density from the centre outwards; e, a pneumite with a polar axis $y\,z$, possessing vibrational influences unequal in different directions. f may show diagrammatically the expansion of the pneumite by heat, p being the limit of expansion by increment of temperature up to the critical point, p' a sudden expansion at the critical point to form a gas. A special pneumite may be adapted to take one form of electricity + or −, so that it can only combine with another

* 'Fluids,' p. 10.

like pneumite under decrease of temperature through the intervention or inclusion of a pneumite of a different sign.

33. The centre of the pneumite may be in one sense a universal form of gravitative matter (*gravite*), or this may be an element of it, upon which alone the amount of gravitation and cohesion depends, while still possessing other affinities. The resistance to combination may depend upon the rigidity or depth of elastic coating of the pneumite, the limiting extreme surface being, however, a constant of perfect elasticity. In the combination of two or more pneumites at equal temperature, g, fig. 3, may represent the

Fig. 3.

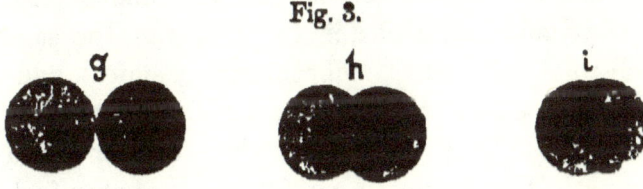

gaseous state at the nearest approach of two pneumites, h this state in combination, i two inseparable pneumites at the same gaseous temperature. With the like factors to those described above as the special characters of separate pneumites, the final compound atom may possess any observable qualification due to its composition. In the same manner as the apparently similar cell in organic life possesses the elements of the functions of the organ to which it belongs.

Taken in this manner, a free atom may be considered to resemble in a certain degree the physical state of an organic being, possessing potentialities which are active only when it is endowed with the vital force of heat or electricity, sufficient to produce the fluid state; but which cease when this influence is withdrawn or dissipated. So that the atom remains encysted, as it were, in a dormant state, when it forms part of a solid mass, outwardly sensitive only to the properties of cohesion and static equilibrium, so as to be affected by temperature and electrical excitation only to a limited degree.

If there is a distinct pneumite of entirely discordant vibrational period with other pneumites, it will form a permanently dissociated system—give a single line in the spectrum—have no power of association or absorption with other matter, and be impossible of condensation from the pneuma state. The pneumite is assumed to be the prime mover of light vibration, which may be communicated through ether or otherwise to a distance.

The mode of construction proposed above for the initial units of matter appears to me to be simpler than that of assuming any single factor to possess at the same time many distinct properties, vibrational, chemical, and other. Indeed, it is easier to conceive 10,000 varieties of such structurally simple distinct units than one endowed individually with the many properties of the chemical atom, needing Clerk-Maxwell's little demon to direct it.

34. In the general pneuma system here proposed as being formed of specialized pneumites, although the pneuma may be invisible itself when widely dispersed in space, it may form in great bulk a density system by mutual attraction, and even possess a certain degree of absorbent or refractive power upon light passing through it, as before suggested for attenuated nebulæ. It would evidently in all cases be placed exterior to the denser nebulæ before its condensation thereto, which may meet Sir Wm. Herschel's suggestion that a nebulous appearance of stars may sometimes be caused by their shining through an attenuated medium*.

35. *Chemical Action in relation to a Pneuma System.*— Upon the data just proposed, if we assume that our solar pneuma system originally extended to the primitive radius of other star systems then in condensation, we may imagine

* Phil. Trans. 1811, p. 359.

the synchronism of rates of vibration of certain classes of dissociated elements or *pneumites* would, by equal unity or multiple unity of vibrational period, promote association in groups to form what we recognize as the chemical atom, which may be compared roughly to a chord in music in relation to its separate notes. If the atom is complex, a better comparison would be to the notes of an organ tuned to a certain pitch, sometimes with omission of certain notes and with diatonic intervals between others. The number of original pneumites in any chemical atom may possibly be represented by the number of lines in the intense spark spectrum, and the strength of these lines gives the relative quantity of special pneumites of one vibrational period in the atom. In this form of association it follows that any *pneumite group* or, as we should term it, *chemical atom* is that system of pneumites which can associate with the least friction among themselves to form more concrete matter, but which would eject or exclude other near pneumites moving at a discordant rate of vibration period. The excluded pneumites might again unite in simple groups of coincident vibration to form other distinct chemical matter. The atoms or matter formed from a pneuma system would remain hereafter invariable and initial to that particular system in which they were formed. If separated temporarily by heat or electrical action they would remain near, at positions less frictional than that of entering other groups, and therefore would again unite from the same matter into the same atomic forms.

36. In the above construction it will be seen that a lower temperature or diminished electrical excitation, by diminishing repulsion or its equivalent, will permit approach and constantly promote association, so that this association will give a mean vibrational period to two or more groups of pneumites of multiple accordant vibration period. Therefore a new vibrational position in relation to the spectrum for the first or earliest grouping. A still lower temperature may

produce another less divided molecular system, leaving only a single group of molecules sufficiently active synchronously for spectroscopic observation, the vibrations sinking by loss of temperature or electrical excitation to a perfect molecular concentrated condition of restricted vibration, moving below a light-giving period of energy.

37. It will be seen also from this proposition that by vibrational unity an element may possibly be formed minus certain pneumites, or with a greater or less number; as, for instance, the spectrum of another star system, as suggested by Prof. Lockyer, may resemble our own in certain spectral lines, through being formed of pneumite groups or chemical atoms from the prevailing pneuma slightly different from the solar constituents. The sun or a star may possess a metal that may resemble iron in many particulars and yet not be in all chemical properties exactly like our iron, from the difference of pneumite composition shown in the quantity of special pneumites which constituted the pneuma from which it was formed; but once formed it would be universal to the special system. Any line in a star may be omitted, or other lines added in an approximately like chemical element. Certain pneumites of accordant period may combine in several groups, as, for instance, certain hydrogen pneumites of the C, F, G groups may be present also in nitrogen, only slightly displaced from a normal position, or even in a kind of duplicate motive action giving two lines instead of one, through collateral or rotative vibrational influences of other combined pneumites special to the nitrogen or the hydrogen atom. The sensitiveness of any special group by its collateral internal vibrational freedom would produce sufficient vibrational amplitude for spectroscopic observation, whereas another more restricted motive group, as before stated, would fail in this, and therefore be invisible.

38. *Cohesion of Matter.*—Further, upon the above premises, there must be a unity of vibration-period for the

concrete chemical atom derived from the mean momentum of the associated periods of the pneumites of which it was formed, and the possibility of a like unity of atom and atom to form the molecule. Therefore the possibility of a unity of *space-motion* between like chemical atom and atom and molecule and molecule, and thereby a possibility of near approach when in like phase causing such atoms or molecules to interlock as it were and form denser matter, or to form a local cohesive system, to which surrounding free atoms or molecules would be drawn and adhere by central affinity or gravitational influences to form mass or visible material. In this manner in the nebular system, as the denser masses of associated matter, atoms, or molecules approach, through the influence of what we may term internal cohesion to gravitation-centres of attraction, the sum of these motive systems of discordant vibrational period, or those constituted of lighter vibrational momentum or of higher elasticity through simplicity of pneumite-composition, forming altogether what we recognize as lighter or more repulsive matter, would be displaced to the exterior parts from gravitation-centres, as hydrogen and helium are about the stars and visible nebular systems. The motive energy lost by the concrete association of pneumites to atoms, and atoms to masses may be transferred to surrounding space. This energy, being dissipated through the outer attenuated pneuma, or the ether, will appear as tangible heat or visible light, in its encounter with any exterior material body *.

39. It is possible that an atomic system may be formed of separate pneumites not of *absolutely* concordant vibrational period. In this case the atoms formed of pneumites of perfectly concordant period will be stable, and in cohesion also stable, as gold. Atoms formed of pneumites partly of slightly discordant period would be unstable, and the same in cohesion. Atoms formed within the greatest possible limits

* Appendix A.

of discordant period would be explosive. Atoms in chemical combination with other atoms of slightly discordant period would be also explosive, although they might be *per se* as regards pneumite composition of concordant period. On the other hand, atoms may have concordant period, although their pneumite composition may be partially discordant with other atoms. Matter formed of such concordant atoms with other atoms would have a tendency to associate in chemical composition or alloy, as nickel and cobalt, the two groups of platinum metals, yttrium and didymium, &c.

40. In a vast pneuma system supposed to be in a state of elastic agitation through heat influences and possibly electrical excitation, which may afterwards form a solar system, we may assume that the primitive pneumites move in multiple vibrational periods, x, $3x$, $25x$, ... r, or y, $5y$, $9y$, ... r. Since x or y may represent any number of vibrations in unit of time, these pneumites would be ready to unite into atomic systems. Therefore we may assume these concordant atoms would form the bulk of cosmic materials. For instance, iron with its great number of spectral lines may be the predominant material. This is consistent with its predominance in the meteoric matter which falls to the earth, condensed from the universal pneuma. The last pneumites to unite would be those at the limits of the possible approximately vibrational period that would be capable of forming atomic systems. These would bear some relation to the tempered notes $c\sharp$, $d\flat$ in pianoforte tuning in relation to f, g, a. Or as αx, βx, γx, $3\alpha x$, $5\beta x$, $27\gamma x$ ... r, where α, β, γ represent the *beat* periods only of the vibrational time of x. Matter formed of such atoms would be possibly rare in a cosmic system, as the greater number of pneumites would be selected to unite in the permanent atom-groups of simple multiple period. Such complex atoms in $\alpha \beta \gamma$ periods would be possibly open to dissociation or to form separately material bodies, α, 3α, 11α, ... r, or β, 5β, 6β, ... r. Under such

refined systems of analysis as that of fractionation by Crookes, therefore such factors of matter would represent the meta-elements of that philosopher *.

41. *Observation of Astronomical Nebulæ.*—The theory of the nebular condition derived from observation of the bluish or greenish unresolvable nebulæ, having regard to the facts revealed by experimental science, appears to be best explained by assuming these nebulæ to be purely gaseous. The mode of condensation which I would suggest, which renders these masses visible to their extreme outline, is purely chemical in the form herein proposed and takes place within the exterior surface of the nebula only. The nebula may or may not possess a central incandescent nucleus. The chemical action results in the degradation of the pneuma to a gas or a nebula, under which action free electricity is developed.

42. In the early purely pneuma state all the elements may combine by mixture, but still remain separate distinct pneumites. The condensation of pneuma to the nebular state causes matter to fall towards the centre through cohesion, where a secondary central system of illumination through heat may be formed by gravity, leaving an attenuated atmosphere of the lighter, more permanent gases in exterior position, as before stated. The effect of the superficial condensation is to develop electricity, just as the condensation of water-vapour in clouds develops it in our atmosphere. Therefore, if we could see a nebula quite near, it would within its surface be continually sparkling or possibly be suffused with dispersed flashes of lightning. These sparkles or flashes would not occur in the extreme limiting surface, which would possibly be of hydrogen and helium in a high state of diffusion, but in a somewhat lower stratum, where possibly air or nitrogen would form a denser layer, probably united with aqueous vapour as in our atmosphere. These flashes are assumed to produce the bright lines of nitrogen

* See Wm. Crookes's important Address to the Chemical Society, 1888.

or helium occasionally seen in the spectroscope. The exterior hydrogen and helium, in greater tension than that of a Geissler's tube, becomes electrically excited and forms also an insulator to the internal matter which is under intense chemical action, or under vivid electrization through the chemical action; so that we may consider an apparent nebula as the seat of an auroral display. This superficial action of forces, producing light-vibrations, would detract little energy from a large volume of matter in a nebular system, whereas the radiation of intense heat from the entire system sufficient to produce the light we observe would detract much, and still not account for the spectroscopic phenomena.

43. *Condition of the Sun.*—As the sun is evidently at a temperature above that of dissociation of the chemical elements, it must be represented by a permanent gas, the gravitation of the mass causing great pressure in its central parts. Under such conditions there would still be the tendency for the exterior dissociated pneumites to unite into equal or multiple vibrational systems, although the elasticity or repulsive action of the surface of the pneumite could never at its temperature be overcome by pressure so as to allow of a very dense system being formed. Probably the constant tendency of the pneuma to unite into a density system in vibrational unity is the cause of intense chemical action, development of heat, and electricity. Through the internal friction of the heat vibrations of the system a liquid density-system may never be approached under present conditions. The temperature of the sun at present, no doubt, dissociates all elements, and projects the dissociates (*pneumites*) from its surface as gas where the pressure is reduced, but such pneumites cannot maintain a permanent state under the open radiation of the sun's surface. They therefore condense to nebulæ and obstruct his direct rays, still moving at their own vibrational period, and become light-absorbing elements. It is probable that very refractory matter projected locally

outwards from the sun's surface may not reach the surface of the photosphere, but that it condenses to nebula through radiation at a lower depth, producing the deeper surface of the sun-spots. These spots may, in this case, be possibly found to be formed of a close series of absorption-lines due to very refractory matter, which may be observable if the central light is sufficient to pass through them to show the inter-vibrational intervals. Therefore it is possible that the more refractory matters invisible in the solar spectrum of the photosphere may hereafter, with refined analysis, be found in the spots.

44. In assuming the sun to have been condensed from an extensive nebula which extended originally much beyond the orbit of Neptune, it would appear at first thought that the lighter, less refractory matter in the gaseous state would remain at an exterior position. This has reasonably been proposed to account for the smaller densities of the superior planets. Upon this argument the presence of hydrogen near the surface of the present sun appears to be anomalous. There are, however, two reasons why it should appear about the sun :—(1) The sun derived its matter from all directions. Tangential motion assisted in prevention of approach of lighter matter in the planetary plane where matter was placed most perfectly in motive equilibrium with gravitation. Whereas above the solar poles, where all matter must approach without tangential resistance, hydrogen would also ultimately approach when all the more refractory matter had condensed. (2) Within the pneuma system proposed the pneumites of hydrogen possibly possess perfect vibrational concord with those of a dense metal, say, palladium, and therefore condense to vapour with it, but at the intense heat upon the present sun-surface the hydrogen is constantly excluded, although it remains permanently about the sun's surface by its gravitation.

CHAPTER III.

FORMATION OF STELLAR AND SOLAR SYSTEMS.

45. *Formation of Nebulæ and Stellar Systems from the original Pneuma System in the Milky Way.*—Under the conditions suggested in the last Chapter we may assume that the whole system of the Milky Way formed an immense pneuma moving in slow rotation, the volume of which included the original places of the matter which surrounded and formed all the stars of the system. Such a system, of the volume the premises infer, could not be maintained static against gravity unless its pneuma was in a highly heated or motive state. In the interior of the system heat would be best conserved from radiation, therefore the exterior parts of the system alone could radiate heat freely into space to condense this pneuma into the nebular state. Taken in this way, the parts of the entire pneuma upon which separate nebular systems could condense by radiation into nebular star systems would be at first only relatively near the exterior limits of the system. The outer condensing parts of the system would, on account of distance, be held only lightly to the gravitation centre of inertia of the entire mass, and there would be present certain elements of tangential action from the rotation. The nebula would therefore experience effective resistance in falling towards the centre from this cause and from the motive elasticity of the interior highly heated parts, which would become by condensation denser than the exterior parts. Under these conditions the early formed

FORMATION OF STELLAR AND SOLAR SYSTEMS. 39

nebula must either float as it were within the surface of the more central highly heated pneuma or be condensed into its mass. In either of these cases the exterior surface of the system becomes at the time of its condensation superficially a denser system than before. Radiation of heat into space being constant, the surface condensation would remain constant, and as the interior elasticity would be maintained by the central heat, there must therefore be necessarily a certain amount of tensile strain upon the exterior matter under condensation produced by the decrease of its volume by shrinkage from radiation. Thus there would be a tendency for the exterior condensing matter of the pneuma system to separate into detached parts or nebular systems formed by the condensation of limited volumes of exterior pneuma contracting upon themselves. These condensations may be conceived to separate the outer pneuma system by tension, something after the manner that basaltic columns in formation are separated upon an equal gravitation plane from their uniform matrix. After separation such unit nebular masses would gradually draw their matter together into higher density systems, and finally by gravitation into globular volumes, the matter gradually further increasing in density in time towards their own centres.

46. The condensing systems described above, if seen from a great distance, would at an early stage resemble systems of nebulous stars spread out upon a thinner nebular base. As soon as the exterior parts of the nebular system formed a more condensed or separate concentric nebular star system, these nebulous stars would further contract upon themselves, and then the interspaces thereby produced would permit the free radiation of heat from the more interior parts of the entire system of the pneuma, so that upon the same conditions another concentric stratum of condensations more interior than the first would again form a nebular star system, and this again another in like manner still more interior,

until the whole system of pneuma here considered was condensing through radiation of heat to nebulæ and finally into stellar systems under initial gravitation with intense chemical action. Upon this principle the pneuma would be separated into unit nebular star systems, and finally into individual stars each radiating heat proportional to its condensation into universal space, as it does at the present time in our Milky Way.

Certain local systems as here defined possessed of great central density, if seen from remote distance, would resemble in the whole or in partly advanced stages of condensation some of the globular or lenticular clusters such as Messier 3, 13, 15, 53, 92, and others.

47. *Suggested Motive conditions of the Original Pneuma of the Milky Way.*—If this pneuma was originally motive by rotation about a centre of inertia or otherwise, as just proposed, its separate units, detached by local condensation, would be also motive by continuity of the original momentum, which in a perfectly fluid system would be diverted in any direction that at the time presented the least resistance to its motivity. There is not, however, in the Milky Way the perfect symmetry necessary to infer an original uniform rotative pneuma formed under simple condensation. Our Milky Way is apparently an immense flattish bifurcating plane of great depth formed of stars unequally distributed. Another mode of formation may be suggested which, although entirely hypothetical, or may be fanciful, is not altogether inconsistent with the wonderful variations in the forms which exist at the present time in visible nebulæ. These proposed conditions may also have partly induced the separation of the pneuma system into unit star systems, and are quite consistent with the motion of matter upon the nebular hypothesis of Descartes, which is so ably supported by M. Faye.

48. Assuming that there were vast pneuma systems moving originally freely in space, the tendency of such separate

systems in isolation under the influence of gravitation would be to form spheroidal systems. The flattish form of our Milky Way may possibly indicate that two such spheroidal systems of immense volume at an early period may have drifted together. Such a collision as would be produced at the meeting-surface would form at once a relatively superior density plane where the matter of the two systems would be united, and henceforth the centre of gravity of the two systems would be changed to the centre of meeting-plane of the two systems. The pneuma systems being in the highest degree elastic, would continue the momentum of their original projection of mass in direction normal to the plane of contact for many millions of years after impact, continuing also to compress and flatten out the matter of the original meeting-plane. The pressure in this plane would be increased independently of original momentum by the action of the mutual gravitation of the two systems in approaching to unity throughout the meeting-surface. One effect of the compression at the meeting-surface described above would be the development of intense heat and electrification. The electrical excitation would diffuse itself through the entire mass, possibly rendering it luminous to its extreme limits.

49. The result of such a collision after n millions of years would be the formation of a vast lenticular pneuma more intensely heated in its central plane, which would be more dense from compression and gravitation than the surrounding parts. If the two parts did not entirely combine, a bifurcating system might be formed. The whole system would under any condition possess a momentum compounded of the original momenta of the two pneuma systems from which it was formed, but in its separate parts it would possess motivities consistent with the conditions brought about by the collision and reflective reactions of parts of the perfectly fluid matter of the pneuma systems. Without such a collision as herein depicted it is possible that the condensation of the stellar

pneuma systems of the parts of the Milky Way would have been much slower than they were, so as possibly not to have advanced at the present time beyond the condition of what may be the present state of outer matter about the galactic poles.

50. *Cyclone-inducing conditions.*—Assuming the pneuma to be a most perfect fluid and elastic system of matter, upon the meeting of two volumes of such matter, independently of any initial rotation it might possess, must have moved under pressure at the meeting-plane in every or any direction which at the time offered the least resistance to the continuity of its initial momentum. Therefore, in the meeting of two fluid systems as proposed above, the direction for continuity of motion of the outflow of these systems from the contact plane, where the pressure would be greatest, being supported by the momentum of the following parts, would drive the compressed motive pneuma into directions normal to the pressure. Under the conditions given the outflow would be along the plane of contact and outwards in every direction from the centre of greatest compression, thus projecting the fluid into a current plane. Such a motive plane or current, as I have shown by many experiments in my work on 'The Motion of Fluids' *, engenders motive whirls in the contiguous parts of the surrounding fluid, this form offering the least frictional resistance to continuity of motion of a fluid moving within a like fluid. These whirls, if formed on the principles suggested, would in their turn engender other friction-whirls exterior to themselves, until the entire system became one of complete eddying motion, as it is termed.

51. As this form of fluid motion, originally investigated by myself, is not very generally known, it may be as well to demonstrate this hypothesis by giving an illustration which I take from an experiment in my work on the Motion of Fluids,

---

* 'Fluids,' pp. 227 and 310.

FORMATION OF STELLAR AND SOLAR SYSTEMS.

in which the resistance to a flowing fluid in a plane of pressure is applicable to the conditions suggested above. Although the experiment referred to is on a very small scale, I have shown that motive systems of fluid are proportional and irrespective of scale, so that the same principles extend to the surface motions of the North Atlantic Ocean as hold good in the small experiment I describe * :—" Take two plates of perfectly clean glass, say about six inches square, and firmly cement a border of card about $\frac{1}{18}$ of an inch in thickness round three sides of one of the plates with marine glue. Then cement the second plate to the first. By this means a very thin waterproof trough will be formed, open at the top. If we now fill this with clean water and place it upright in a groove in a piece of stout wood, the experimental apparatus will be complete.

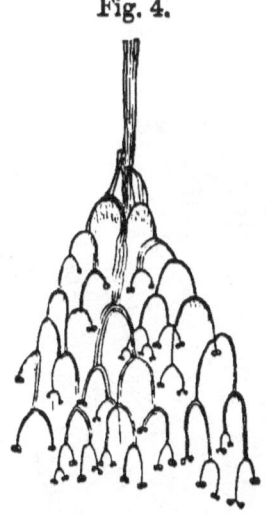

Fig. 4.

" To observe the desired effects, the trough should be placed before a window to transmit light through it, when the following phenomena of original fluid projection may be observed by the projected slightly heavier liquid diffusing

* 'Fluids,' p. 366, *i.*

itself in a current, through the constant force of pressure of its small excess of gravity.

"Take a pen full of ink and place this gently upon the surface of the water in the trough. The ink as it descends slowly, by excess of gravity over the water, will be found to divide constantly upon the resistance which opposes its direct projection, the divided parts having insufficient momentum of projection to move the lateral fluid tangentially to produce extensive lateral whirls. After the first division of the descending fluid the divided parts again divide, and these again, so that by this constant division and after subdivision the projected fluid takes a tree-like form, the terminals being in spiral rotation. The illustration (fig. 4) was taken by transmitted light, exactly following the outlines of the projection of ink in the thin trough described; it is therefore represented of the observed size" *. In this experiment the sides of the trough are assumed to represent the pressure planes.

Fig. 5.

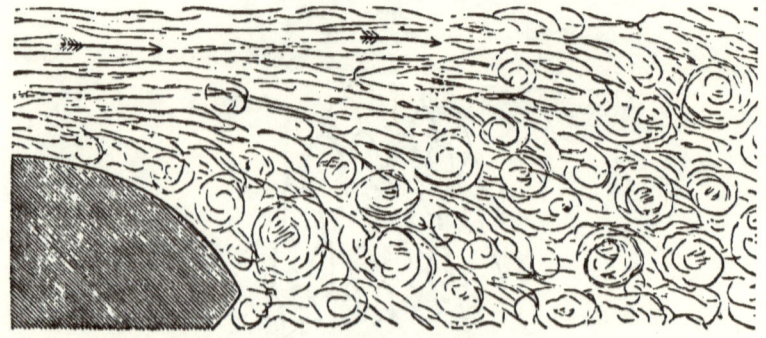

52. The principle of lateral diffusion of whirls is seen contiguous to any flowing stream in lateral water: the sketch (fig. 5) was taken from the flow of water through an arch of London Bridge.

* 'Fluids,' p. 314; *id.* p. 310.

In this matter we might consider the separate motive cyclonic volumes of nebulæ in a very attenuated system to be an early form of projected matter inducing rotation in stellar formations, as in the theory propounded by Descartes and so ably supported by M. Faye. These motions must, however, be considered to take place strictly within the pneuma, and would not be the cause of its condensation or separation into stellar systems, the conditions of which as effects of radiation have been already discussed, § 45.

53. *Formation of Spiral Nebulæ and Stellar Systems.*— Taking any separate unit system of pneuma as a part of the system just proposed or any other, we will assume this to be in uniform angular revolution, its exterior parts moving with a velocity below that which would project them in an orbit. The outer matter of the system being assumed to condense through radiation, then particle would approach particle as the near parts of vapour in the atmosphere do to form rain-drops. Such particles or drops would follow in the general drift of the residual vapour of other matter that could condense only at a lower temperature as rain-drops fall through air. Under these conditions the earliest condensed outer matter would drift as a discrete system in showers or congregations or concatenations under the surrounding resistance in spiral lines towards the centre of the system with increasing velocity owing to acceleration by gravity*. Such congregations of material discrete particles under the influence of mutual local attractions would present a feathery or flocculent appearance on the breaking up of the outer pneuma of the system. They would possibly be illuminated by friction on electrical excitation. As soon as the isolated flocculi drifted in spiral paths sunward or to the star-centre, other surrounding condensing matter would follow in the same direction by initial gravitative influences, and form currents

---

* Newton's 'Principia,' Lib. ii. prop. xv.

in certain positions piercing through the minor resistances of the lighter interior gaseous matter which condenses only at a lower temperature. In this manner a condensing nebular system moving in spiral lines sunward would appear, if seen from a distance, to be a perfectly stationary system, although its condensation might be progressing at a rapid rate, as the same position of spiral stream-lines of drifting flocculi would be constantly conserved locally. This may be the most general form of nebular condensation.

54. *Solar-Planetary inducing conditions in Spiral Nebulæ.*—If the original tangential motion of a pneuma system, condensing with acceleration of its flocculi through gravitation into a spiral nebular system, attained sufficient velocity through gravity to maintain the flocculi, drifting as above proposed, at a certain position in an orbit exterior to the central system of the pneuma, this matter although in revolution would remain in equilibrium at a certain distance from the central sun or star, just as though it floated on frictionless water. It would thus draw its separate parts together within certain limits by its own internal initial gravitation and the drift of following parts, so as to form a separate nebular condensation or zone exterior to the centre, which thereby ultimately would become in a condition to form a planet. The direction of rotation of such an orbit-zone or planet will be discussed hereafter. In the above we may see clearly the possibility of a partially discrete system of condensation of exterior matter becoming a factor of solar nebular planetary formation.

Upon the conditions proposed in the above and the preceding paragraphs, we may suggest that, in the condensation of the volume of pneuma necessary to form a solar-planetary system, a nebular system may be formed at a certain period by central condensation, which will maintain the nebular condition about the centre, but that this may be surrounded at an early stage by discrete condensations falling thereto, formed by local exterior condensations due to excess of ratiation of heat from the periphery of the system, so that the purely nebular condition for the entire system may appear in some cases in an early stage only. In this construction we have certain factors of the nebular theory of Herschel and Laplace in conjunction with the discrete theory of Kant and Faye, which may be possibly observed active in such nebulæ as that of Messier 51 in Canum Venaticorum, Plate II. *f*.

55. *The Permanency of separate Star Systems after formation.*—If we take gravitation to be a force active at all distances of centres of masses according to a given law, its motive energy being in inverse proportion to the squares of distances, and if we conceive condensations to be originally nearly equally distributed within the limits of a unit system ; then all these aggregate or separate systems of matter, although originally in perfect equilibrium, would drift, though slowly, towards one another.

56. We may, however, assume other conditions to prevail under which the permanency of separate stellar systems as local aggregates, when once formed, might be maintained. We may take gravitation to be inactive beyond a certain radius as a possible condition under which stars once formed would remain permanently fixed in space with relation to one another. It is, however, more cogent—as the law of gravitation has been found to hold exactly under every test of experimental observation within the limited though extensive areas circumscribed by the orbits of double and multiple stars of long period—to assume that this force follows the law of inverse squares for all distances. Under this condition a stellar system may nevertheless be permanent if we assume a universal revolution of stars about a common centre of inertia, as originally proposed by Wright and afterwards by Sir William Herschel, the principles of which are exemplified in our own planetary system. This does not, however,

necessarily suggest a unit centre of revolution for all stars, or even for those which compose the Milky Way. The cyclonic systems just proposed would at least be inconsistent with this. There is space enough for many centralized gravitation systems within the more universal system of the Milky Way, in which the stars are placed at distances from one another far too great to render gravitation active upon the members of any separate system beyond producing slight variations in the forms of the stellar orbits surrounding the nearest or strongest central attraction; although there may be even over and beyond this a general drift or circulation of the entire system or any part of it in which the separate systems here considered are embraced.

57. The influence of the knowledge of our solar-planetary system acts so powerfully upon the mind, that it is difficult to conceive the action of gravitation upon cosmic bodies otherwise than with certain elements of tangential motion by which they move symmetrically in a plane about a centre of inertia. In the wide distribution of stars in all directions this may not be universal. It is quite possible and consistent with an original fluid state, that stars may not always form part of a closed system drifting in a circular or elliptic orbit. Such orbits may be contorted in any way by surrounding influences and still retain the elements of original sidereal momentum. The evidence of the variety of form taken by gravitative matter in the distribution of stars in star clusters and in some nebulæ appears to show that stars and matter move sometimes in a system of stream-lines, following one another in a form we may term *pearling* (as pearls of dew). This applies to the actual appearance only, not to the mode of formation. Sometimes these pearls or stars drift in spirals or cyclones, in like manner to the matter that forms the system of bivolutes in M. 51 Canum Venaticorum, M. 74 Piscium, I I. 168 Ursæ Majoris, as shown in Dr. Roberts's beautiful photographs. We can but conceive that there must never-

theless be beyond this throughout all matter, if gravitation be universal, a tendency to approach of mass to mass which can only be resisted by a motion in discrete bodies in which there are certain elements of tangential direction, and this must ultimately tend to cause the whole of any system of matter to approach a symmetrical form of orbital motion about its centre of inertia, which may include the orbital or other motions of its separate parts, as in the case of the sun and planets; but the time necessary for such a complete formation in the universe would appear to be so far infinite that it may never nearly have approached completion in any large division of the stellar system.

58. *Distances under which Gravitation may be active.*—It may be difficult in some cases to imagine the disturbing influence of gravitation at so great a distance as we know matter to exist in stellar space, as inferred above. But, as it is proposed that the action of gravitation is unlimited in space, acting upon free bodies according to its law of acceleration, then a small attraction upon a very distant free body, where the general composition of motions permits approach, will produce a great velocity in a period which may be relatively short for past time. Assume any free body, at a very distant period, to be moving towards our sun with a velocity of one centimetre a year, which we may take as a quite impalpable motion in space; still in one million years, if the body remain in relative position towards the sun as regards exterior influences, it will have attained a velocity of approach of ten kilometres a year; and this velocity would increase with time in like ratio, until in n millions of years, some small fraction of the past, it would fall towards the sun with velocity equal to or greater than the highest ever observed within our system.

59. *Suggested action of Gravity in time in the formation of Circular Orbits.*—If gravity is instantaneous for all distances, an orbit once formed must remain constant in relation to its

focus for all time if it moves in a perfectly frictionless medium. But as the orbits of certain planets and satellites are so nearly circular, it would appear that there must be some reason for this in a circle-inducing quality as a property of the forces by which these bodies are directed. There is no doubt that in the condensation of a spheroidal gaseous system of nearly the diameter of a planetary orbit, as in the theory of Laplace, that circular or elliptical revolution might be brought about by the resistance of the gaseous central matter to change, owing to uniform angular velocity in all its parts. This will be again considered. But a second factor of resistance might reasonably be derived from a time element in the action of gravity in a manner also originally proposed by Laplace. In this proposition, as time is infinite, small constant forces active in the past may have produced great effects by the present time.

60. It becomes, therefore, more rational, whatever we take space to be, vacuum or ether, to assume gravity acting through this space in time, say, with the velocity of light, or with much greater velocity. It is not in this case probable that any time element of attraction could be detected if the distance remained approximately constant, as we may easily imagine the action of gravity to be *induced* *, and that after induction it remains a constant for constant distance. The change of position in approach or recession may produce a change in the gravitation force of induction, and this may take time. Thus, as an instance, in the approach of a comet to perihelion under deferred increasing factors of gravity, it would arrive somewhat behind its time, but as the attraction would still act in like inverse ratio after it had passed perihelion, it would retard and deflect the body if originally projected in a parabolic orbit into an elliptic orbit after this passage. If the body moved in an elliptic orbit, the action

* Appendix A.

due to a time factor in gravitation would decrease the eccentricity at every perihelion passage, and increase the time of revolution in approaching constantly to a more nearly circular orbit. The circle being the greatest inscribed area due to a given mean rate of tangential motion, therefore we may imagine it the state of least internal resistance and of the ultimate motive equilibrium for the orbit of a celestial body in relation to a single centre.

CHAPTER IV.

STELLAR AND SOLAR CONDENSATION.—FORMATION OF ORBITS.

61. *Separate Stellar and Planetary inducing Conditions in the Nebulæ, according to the amount of Rotation.*—Under the conditions already discussed, assuming that any isolated volume of pneuma may form a part of the universe, we may take it as probable that it will form a system of matter which will contract in volume by radiation directly under the influence of gravity until it forms a central gaseous sphere or spheroid. This, if of sufficient size, would form a nebulous star in an early incandescent state of condensation. We may assume that such a system if it has no rotation, or if its rotation is slow, resulting either from the original motion impressed upon the nebula or from matter gravitating towards it from all directions with nearly equal momentum, so that from this reason the combined orbit and mass of the motive parts by the simple action of gravitation produce a globular system in the nebulous state,—such a system would form a single isolated star moving or stationary in space.

62. On the other hand, if the special nebular system considered is in rotation or in cyclonic motion derived from this, matter would approach the centre with much greater facility in every direction other than in a plane extended from its equator, where the tangential impulse of surrounding matter would produce the greatest resistance to the centralizing action of gravitation. So that in this case a star might be

formed of all surrounding matter except that about the plane of its equator. This system at a certain stage of condensation would therefore appear, if viewed from a distance in a direction nearly normal to its poles, as a planetary nebula surrounded equatorially by more attenuated matter. Such a nebula would be in outline of oblate spheroidal form if the equatorial nebular matter were evenly distributed about the equatorial plane. (Plate II. *a*, Great Nebula in Andromeda; *b*, Messier 32; *c*, ♅ I. 200; *d*, M. 81.)

63. The spheroidal form of such a nebulous condensation might be modified to any extent by unequal distribution of the nebula about its focus into a spiral, lenticular, or discoidal shape, its parts being still held with equal permanence by their tangential impulses. Either of the above-described systems, if in equilibrium about its centre with its peripheral parts moving with sufficient tangential velocity to maintain an orbital position, would be ready to separate its peripheral matter into ring-zones or other separate motive systems upon further central condensation, and become finally in a state to form a solar system such as our own with planets of greater or smaller mass *inter alia* moving in orbits about its centre, the conditions for the formation of which will be considered presently.

64. *Limits of a Solar-planetary-cometary System.*—We assume that after separation of any complete pneuma or nebular system from the universal pneuma, the initial action of gravity within this separate system would immediately commence to form a central condensation of greater density, as before proposed, which would react upon surrounding matter in proportion to its mass and inversely to the square of its distance from any part of the surrounding widely distributed matter.

Suppose an original system of pneuma in the remote past taking one direction only, extending from S to S', fig. 6, and that two equal centres of condensation, S and S', were after long periods slowly formed as star systems upon principles just discussed. The neutral position at which a particle or mass would be originally equally solicited by the gravitation of both systems would be at X. This would be a point of early condensation from open radiation, and at the same time

Fig. 6.

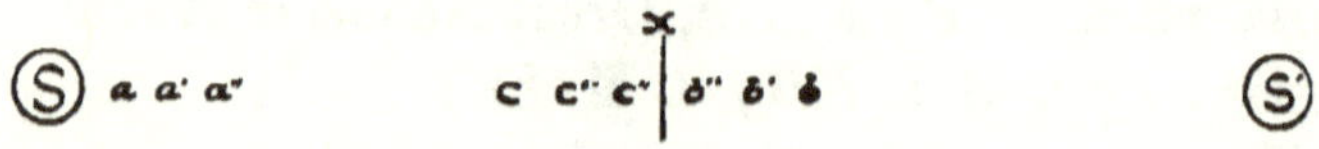

S a a' a'' c c'' c' | δ'' δ' δ S'

one of perfect equilibrium in space. From the equal attraction of S and S' upon the condensed particle, a particle or mass at c'' would be very slightly influenced to move towards S, a particle at c' more influenced, and at c still more so. Thus, assuming gravity universal, this would give attractions to matter to fall towards the centre S from points of rest intermediate between S and S' with velocities inversely proportional to the squares of the distances of S and S respectively active upon it.

65. Like attractions would also hold with respect to the star system S' as regards particles at the relative distances b, b', b''. Therefore all matter from S to X, which we may take as the original radius of our own solar pneuma system, would ultimately form a part of the gravitation system of S with predominant influence over S', and all matter from S' to X would ultimately form a part of the gravitation system of S' with predominant influence over S. Therefore it would be impossible that any body placed between X and S' should be attracted towards S, or in fact that S should form the focus of any orbit of greater aphelion distance than X in this direction. It is further seen upon this hypothesis that cometary or other matter could only move within its orbit at the furthest from the position X in relation to another star S'. If matter does not fall from the positions c, c', c'' directly upon or towards S, there must be a tangential impulse upon

STELLAR AND SOLAR CONDENSATION. 55

such matter in relation to S which directs it into an elliptical orbit. Under any condition, if moved it could attain no greater velocity from gravitation in passing its perihelion about S for projection afterwards than that due to the velocity acquired in falling the distance to S, even in a perfectly frictionless medium. We may therefore note that it is not possible, if gravitation is active for all distances and matter is condensed from a uniform pneuma or nebular state, for any body or comet to reach our sun from space other than that which once formed a part of our own pneuma system.

66. *Action of Gravity on distant Condensations within a Solar System.*—As we may assume that the condensation of the exterior matter of the transparent pneuma would be more rapid from radiation of its original heat into space than that of any more centralized denser matter or gravitation-centre included within it, a considerable volume of outward pneuma might condense on local foci, as before proposed; and these could reach the central space only very slowly from their distance under the slight influence of central gravitation acting upon them. So that a considerable mass might be forming at S, the sun-centre of the gravitation system, by the attractions of a, a', a'' through the centralizing force of S before any effective movement thereto was induced in the distant parts c, c', c'', assumed to be partly under the influence of the star S'. Nevertheless, although the near matter a, a', a'' would experience the centralizing influence of gravity sooner than the distant parts, the interior heat of the system may be assumed to be conserved and supported by that released by the condensation upon the central nebula as before proposed; therefore it would remain a gas for a long time, producing a pressure only about the central sun. The entire condensation of the nebulous sun would be, nevertheless, only in proportion to the radiation of heat from the exterior parts of the nebulous system.

67. With respect to our own solar-planetary system regarded as a rotating condensation under the conditions stated above, we may consider the central nebula formed by pneuma condensation at a point when its extent was represented by N N′, fig. 7, which may be a space including the orbit of Neptune, shown centrally transverse to its plane. Then, if we now limit the extent of directly condensing gaseous sun-forming matter to N N′, the exterior matter represented by C″ C′ C being subject to greater radiation of initial heat and

Fig. 7.

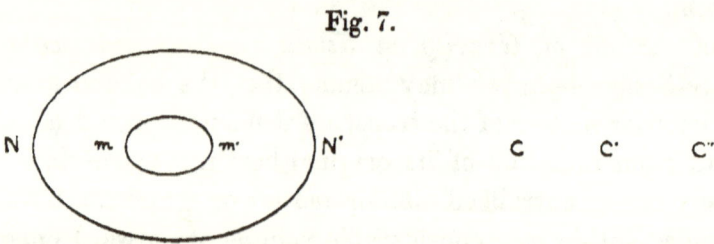

receiving less heat from the condensing sun, these bodies or parts C, C′, C″ might condense upon themselves and draw matter together as before suggested and form separate masses (meteorites), or, if moving in nearly contiguous parts in revolution upon a centre of inertia, form more extensive material systems, flocculi, or *comets*, the probability of which will be discussed hereafter.

If the central nebulous system of N N′ were conservative of initial heat in a certain degree, as here proposed, for a time, the free separate particle or mass C attracted to the central nebula might in time enter the nebula N N′ and be absorbed therein. If it did so, it would become heated by the friction caused by the resistance it would encounter sufficiently for it to become again expanded to gas and afterwards form a part of the central nebulous system, increasing its density thereby. This would not occur, however, without producing a further condensation about the position of entry of the mass and of motion within it, leading to irregularity of constitution of the nebular system. If the included mass projected into the nebula were sufficiently great, it might become an inducing factor of centralizing planetary or satellite aggregation or bring about a disturbance of motive directions of the matter within the system.

68. If we take the condition of matter falling from a position C' further from the centre than that considered above, the gravitation upon this point being less active and the influence of the gravitation towards S' more active, this might not attain a movement in space sufficient to reach the central system until the latter had contracted in volume to a radius represented by $m\ m'$, which we may take for demonstration as the cross section of the sun's nebula when this extended to the orbit of Mercury. In this case, if the eccentricity of the orbit of the matter C' caused its perihelion distance to come within $m\ m'$, the mass C' would, by friction arising from resistance, be retained in the nebula and increase its density, or it would move in spiral lines towards the sun-centre with velocity in proportion to its momentum and gravity to this focus into the resistance of the surrounding matter. If its perihelion distance were greater than the radius m, so as not to come under the sun's attraction sufficiently to be deflected from an elliptical orbit, the body would then move in this orbit constantly for all time and become a permanent planet or comet or meteorite of the system.

It is seen in this that it is only matter endowed with a considerable tangential momentum from any cause that can form a permanent planet, comet, or meteorite of our system, and that the greater mass of condensing matter around the sun at an early period, particularly that which was formed outside the mean planetary orbit plane, would, in all probability, fall directly or indirectly into the sun's nebula and become a factor of sun-formation.

69. If we consider as an extreme case the condition of

another mass C" still more distant from the sun, we may conclude that the small movement induced by gravitation, depending upon the difference of the separate attractions of S and S' (fig. 6), at this distance would not permit it to reach perihelion until the time when the sun had attained its present relatively small volume. In this case, supposing it possessed any tangential momentum, the probability of its falling into the immediate solar nebula would be relatively very small. Such a body would therefore move in an elliptical orbit and become a permanent comet or meteoroid of our system. A condensed mass at a still greater distance from the sun, and nearer S', would not in the past time have reached our sun, so that it may only at a very remote period become what we should recognize as a comet if of sufficient outward volume.

70. In the above construction, in considering exterior matter to be separately condensed and to reach the solar focus in time proportional to its distance, this distant exterior matter would be little affected by the gravitation-figure of the central spheroidal nebulous system assumed to cause its orbit to be drawn toward the nebular plane under conditions originally suggested by Kant. Therefore the exterior condensations of a pneuma system would be projected at an angle to this plane, with the reservation that the tangential momentum would be less proportionally to the increase of angle to this plane, if the original pneuma were ever in uniform rotation, so that the orbits of comets approaching the sun at angles considerably inclined to the mean solar-planetary plane should be more eccentric than those at smaller inclination to this plane, so far as the above stated conditions hold.

71. *Direction of approach to the Sun of matter from the exterior of the original Solar-planetary Nebula.—Formation of Orbits.*—Let A and B, fig. 8, be two stars, or suns, O at the intersection of the lines A B, $d\,d'$ be the point of equilibrium

between these suns, where O would rest in static equilibrium, the gravitation of A and B being equal upon it. Place any number of particles in line at right angles to A B from d to d'. Then gravitation acting upon any of these points

Fig. 8.

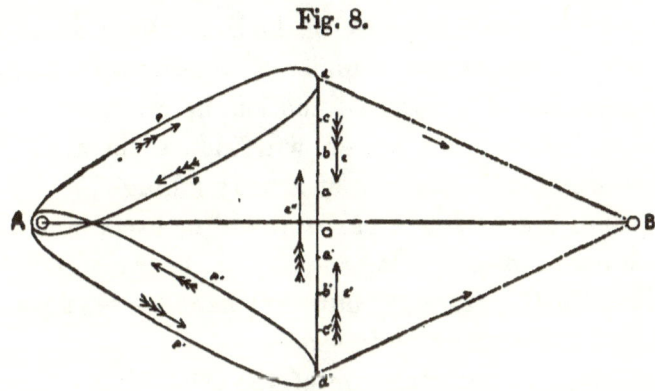

a, b, c, a', b', c', considered as free bodies, would cause them to fall to O, and beyond it if the impulse was sufficient, so that they would oscillate about O constantly in the mean gravitative tangential plane of A and B.

72. If we make A a superior attraction over B, either from nearness or excess of mass, then matter attracted from O would fall in direct line towards A. Any other matter in one of the positions a, b, c, a', b', c', although it would receive less attraction to B than to A, would move in angular direction towards A in composition with the pull of B upon it. Therefore the composition of forces A and B would induce a definite amount of tangential motion upon all the points a, b, c, a', b', c' moving towards A. As the objects represented by points approached A in moving in their orbits, the influence of B would diminish in gravitation proportion; A becoming entirely dominant when the matter of any of the points approached the perihelion of the orbit whose eccentricity was induced by the tangential element B acting upon it.

73. It is also seen by the above diagram that if the masses are distributed at equal distances upon the tangent $d\ d'$ there will be many such masses a, b, c, a', b', c', influenced by the attractions of B in falling towards A, whereas in the direct line O there will be no tangential influence from B. Therefore, the general paths of bodies falling from space from surrounding attraction under a superior attraction, will possess elements of tangential motion in reference to their predominant attractions which will induce them to follow elliptical orbits, and the case of a body falling directly from distant space upon the sun or a star will be exceptional. The principle here shown by diagram applied to two stars or suns will apply equally to any number taken in separate pairs distributed as they may be in space.

74. As regards the direction of the orbit of any free body passing under the influence of the sun's attraction, it will be seen that the amount of tangential motion induced by B upon d' acting in the direction shown by the arrow e' will approximately direct the path of the body in the line $p'\ p'$, in its orbit in a left to right direction at perihelion round the sun and back to its first position. In the same manner, a body projected to A from d with the tangential component induced by B in the direction of the arrow e, will follow the path $p\ p$ in a left to right direction in the plane of its orbit.

75. If we regard any initial motion of the circumscribed system about our sun in the direction of the arrow e'' as an original nebular condition, then we may assume that the influence of B upon matter projected towards A is neutralized at some point, say b, so that b will fall more directly towards the sun A than O, and the scale of distance for projection from this point b for the other points d, c, a, a' will be equal to that previously defined for the point O as regards tangential action in inducing direction of revolution. In this case the whole system of separate attractions as motive forces may be assumed to be displaced by this initial direction of motion,

and the preponderance of direction will be that of the initial tangential motion of the whole system, but not of every element of it.

76. The above scheme, fig. 8, which is assumed to be that of the form of gravitation action between our sun and every near star, may be placed at any angle or direction to the polar axis of the sun where the star appears. If matter in the positions $a, b, c, d, a', b', c', d'$, shown by the diagram, were projected at an early period near the mean planetary plane, the exterior body would fall into the planetary nebula, combine with it, and enter into composition with its motion. This must have been the condition of some of the early comets, assumed to be flocculi condensed within the superior influence of the sun's nebula; and as these condensations must have formed in all directions with regard to the sun, the comets left at the present time in our system must be much fewer or more particularly of much less mass in the mean direction of the sun's equator and the planetary plane, than in the direction of the solar poles. This condition is not, however, absolute for all points, as a near star in the direction of the solar pole or any other direction would as a gravitation centre cause much less matter to fall directly towards our sun than that which would fall in the direction of a more distant star or from an intermediate space. This subject will be taken into consideration further on.

77. In the construction shown in the diagram, fig. 8, it must be clearly observed that unless the entire system is in motion in relation to a gravitation centre, as shown by the arrow e'', the sum of the deflections of the points a, b, c, a', b', c' towards the plane AB would be equal; so that no rotation would be imparted to the sun A or a planetary system connected therewith by the momentum of the united masses of this matter if it was impressed upon the sun's nebula at any period. This does not appear to be clear to some authors who have considered the subject without

proposing clear definitions which, it is hoped, the diagram may supply. Kant suggested that the sum of collisions from exterior matter attracted from all directions towards the sun from space would direct his revolution and that of the planets in one direction as a resultant. Herbert Spencer adopts this view *.

78. M. Faye shows that the mean momentum of all surrounding bodies drawn in direct line by gravitation would not impart any revolution whatever to a central system †. This is herein demonstrated; so that although all comets may be considered as extreme condensations that take one or the other direction of revolution in long elliptical orbits, upon the principles discussed above, yet, if they were retained within a nebula surrounding the sun at perihelion, unless the nebula possessed original rotation, the mean motion produced upon the rotation of this central nebula would be approximately *nil*. Taken in another form, excluding original rotation, we might conclude that the probability is that the mean momentum of all the comets of our system is approximately *nil*, there being possibly as many, or, more exactly, as much exterior matter moving from space in one direction as in the other.

79. *The Formation of a Planetary Plane.* — Under the above-stated condition, during the time that the sun remained an extensive nebula any exterior body entering this nebula at high velocity must have become dissociated and incorporated therewith by the heat engendered in the nebulous matter through friction, its initial momentum being compounded with that of the nebula into which it was projected, as before stated. Therefore, all exterior projections into the sun's nebula will be drawn towards a line passing directly from one sun to another predominant sun as the linear direction of greatest attraction, and upon this principle the

---

* "Nebular Hypothesis," The Westminster Review, 1858.
† 'Sur l'Origine du Monde,' 1885, p. 134.

final mean equatorial plane of revolution of the solar system must have become that of the mean directive influences of attraction of all the near stars or other matter that surrounded it, at the period of its formation, in combination with the momentum in revolution of its original pneuma or nebular system. This does not, however, infer that the sun itself rotates in the mean plane of the original nebula. Its plane of rotation would be largely influenced by the amount of matter that was condensed upon it from any direction and the momentum imparted thereby. This will be considered further on.

80. *Planets formed at the Perihelion of Cometary Orbits.*— Under the condition that the greater part of the matter that was locally condensed, falling sunward from interstellar space within the radius of the sun's superior attraction, would fall into very elliptical orbits, diffused matter would become much more dense near the sun where the perihelia of the projections meet. Therefore, if we conceive the sun to be at any period in a state represented by any of the spheroidal planetary nebulæ, so that the nebular density decreased constantly from the centre of the sun, such a system might continually contract in volume by radiation and yet maintain a similar outward state from exterior pneuma projections thereto.

If the perihelion distance of any such condensation as described above fell within the attenuated nebular system about the sun, its after projection therefrom would be of less velocity, so that it would follow an elliptic orbit of less eccentricity. A second perihelion contact with attenuated matter about the sun would again act in like manner, so that by perihelion resistance matter projected thereto might finally fall into a nearly circular orbit. If the resistance were of a certain value the above conditions would occur at a single perihelion contact. If the resistance at perihelion exceeded the momentum required to maintain a nearly circular orbit, of about perihelion distance, the projected body would, as before stated, drift in spiral lines towards the centre ; or this resistance might be partial, so that it might rest in orbital equilibrium in a certain internal position and become a part of the permanent nebula or planet moving henceforth in a nearly circular orbit. If its perihelion of projection glanced upon the outer surface of the solar nebula, entering it only sufficiently to be deflected from the resistance, the projected body might remain nearly in contact. This must, however, be a special case. The near comets may have been the result of such glancing conditions. The planet Pallas may have been formed under similar conditions from a comet that glanced upon the limit of the solar nebula sufficiently deep to be resisted at its head, until its tail coming forward in the same direction condensed about the head and formed a nebulous planet, to be henceforth projected in an elliptic orbit of small eccentricity. The nebulous planet would afterwards condense to its present state by radiation and initial gravity.

81. In the formation of our solar-planetary system, the fact should never be lost sight of that the entire planetary system is of relatively small mass in comparison with the central solar system, being probably not over 1/700 part, including all planetary, cometary, and meteoric matter. Therefore the parts of the solar-planetary system which attained orbital velocity during the nebular condensation at a distance from central solar nebula, as a resultant of original rotation, acceleration through centralization by gravity, and by the perihelion retention of exterior matter here proposed, must altogether be taken to form but a small part of the mass of the entire system. In the further discussion of the formation of the planets this conception of the subject will always be inferred.

CHAPTER V.

DISCUSSION OF THE MECHANICAL PRINCIPLES UPON WHICH OUR SOLAR-PLANETARY SYSTEM MAY HAVE BEEN FORMED.—SUGGESTED DEMONSTRATIONS OF THE THEORY OF LAPLACE, WITH SOME MODIFICATIONS.—LIMITS OF A COMETARY SYSTEM.

82. *Energy of the Solar System.*—It may be held that Lord Kelvin has shown demonstratively that the energy of the solar system could not, even if it were produced by a discrete condensation of cosmic matter, have been maintained by this form of condensation throughout the narrowest possible limit of past geological time (§ 9) *. Therefore we have no theory heretofore offered of a condensation system by gravity to represent the formation of our solar-planetary system with any reasonable probability other than that of Laplace and Helmholtz. Nevertheless it is not probable that our system was formed by any simple single mechanical effect of the action of forces upon surrounding universal matter, as generally assumed in special theories, but rather that all possible conditions were active that may have conspired to produce the final results. Some of these conditions will be now suggested and discussed.

83. *Asymmetry of our Solar-Planetary System.*—If we suppose our original nebula throughout its entire volume to have been in a uniformly purely gaseous state and of symmetrical

* Trans. R. S. Edinburgh, vol. xxi. p. 66.

form,—as, for instance, that of a spheroid in revolution—we should then, no doubt, if the entire system remained gaseous until all the planets were consecutively condensed at the exterior limit of its nebular equator, according to the theory of Laplace, expect to find the planets in symmetrical order of distance and of mass. In this case, with proportional time-condensation, under the increasing amount of tangential impulse due to centralized condensation into gravitation, which produces the law of orbit, the distances of the planets from the sun and their separate masses would be symmetrically proportional, in accordance with the pull of gravitation and the tangential momentum of the amount of the condensed matter.

84. That our solar system does not possess the above described symmetry is evident from its formation. We have between Jupiter and the earth, particularly, planets in mass and density in no way proportional or symmetrical with others. The planets exterior and interior to the asteroids, taken by themselves alone, have some points of resemblance in density and in magnitude. The exterior planets fairly resemble one another in number of satellites, assuming that Neptune may have several more than the one visible through our telescopes; so that so far as these conditions go we might roughly divide the system of planets into two classes. We might also possibly make a much more important division, in a formative sense by separating them according to the direction of rotation, by which the outer planets Uranus and Neptune would form a class by themselves. Notwithstanding all these minor classified resemblances, there evidently remains beyond this a general want of symmetry in the entire solar-planetary system. Therefore, if we conceive an original uniform gaseous or a generally symmetrical system for our primitive nebula placed under the condition of uniform condensation, according to the theory of Laplace, we must conclude that this system has experienced material modi-

fications during the period of condensation to form our planetary system. This will be now considered by taking at first as groundwork the effects of the condensation of a purely symmetrical nebula, and afterwards suggesting what modifications there may have been in some parts of the system.

85. *A Symmetrical Gaseous Solar-Planetary System.*—We may take our solar system in its nebular state at a certain period to be of the simplest construction shown diagrammatically by fig. 9, which is intended to represent a very oblate spheroid in revolution upon its symmetrical axis, $a\,a'$. We

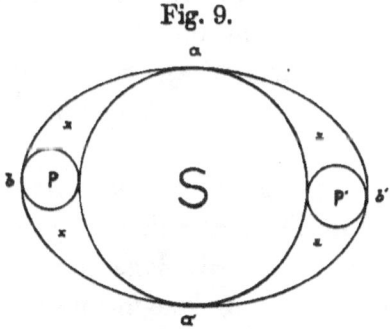

Fig. 9.

may assume that the whole spheroidal surface of the nebula would be radiating heat equally per unit of area into space. The volume of nebula in a gaseous state would therefore be limited at any time by the quantity of matter that could be maintained in this state, by the initial heat of the primitive gaseous system, together with the amount of heat given out from the centre, where there would be a sun-forming condensation.

86. In the above-stated case the important heat-maintaining centre or incipient sun would react through its condensation as a heat-radiator to the surrounding nebula, and disperse its heat due to condensation in proportion inversely to the square of the distance from the incipient sun-surface. The heat-radiants being therefore equal at equal distances from this sun, would tend to maintain a circumscribed *globe* of nebula

in a gaseous state, if its exterior temperature was falling. We may assume that this inscribed imaginary hotter globe, fulfilling the condition of equal radiants, would be so large that its surface could just be inscribed in the oblate nebulous spheroid which we now take for our complete nebula, as shown by the inner circle of the figure.

It is readily seen that matter placed outside the theoretical globe suggested above would be more rapidly lowering its temperature by radiation, from its greater extent of surface in proportion to its depth of volume, than the matter of the inscribed globe. At the same time this external matter would be receiving less heat from the sun-forming centre. The peripheral matter in the general revolution of the system would also possess its highest tangential velocity in the equatorial plane.

The whole of these conditions, upon loss of heat of the system by contraction through radiation, would cause a stress at a certain nearly cylindrical plane parallel to the axis of rotation, wherein internal or sun-forming matter would, by continuous cohesion or gravitation, in the gaseous state, be drawing away from the peripheral zone, which was moving at higher tangential velocity. This zone, by continuity of radiation of its own heat after its detachment, would be induced to contract by condensation upon itself.

87. In this manner, if we assume peripheral matter to be proportionally distributed about the axis of revolution of a spheroidal nebula, the stress-plane would be that of the separation of the peripheral matter, which might then form a detached zone or ring upon the theoretical conditions given by Laplace. The extent or distance of the ring from the sun would depend upon the extent of the globe of centre-tending condensation maintained by the central heat at the time. That is, really upon the contraction at the surface of such a globe as herein imagined, by loss of its heat through radiation.

DISCUSSION OF MECHANICAL PRINCIPLES. 69

In the separated ring or planet-zone, as here suggested, we presume a perfectly equal distribution of matter about the equatorial regions of the oblate spheroidal nebula; but as we take the conditions of radiants from the sun-centre, the same principles would approximately hold if the ring were more or less incomplete or denser in any part. The accident of a perfect ring or system of rings which occurs around Saturn may be conceived possibly to be a unique phenomenon of the nearly equal distribution of cosmic matter about a gravitation-centre.

88. The principles here discussed are a possible explanation of those given by Laplace, under which the nebula forming the exterior planets would be consecutively abandoned. Whether there may have been some modifications of this during the condensation of our own planetary system will be discussed hereafter, but to follow our theme, we will take a uniformly distributed nebular system in revolution as being of spheroidal form, in which case the nebula might continuously go through the same set of changes as consecutive zones or planetary systems were detached from the central sun-system. Thus, assume (fig. 9) $a\,b\,a'\,b'$ to represent diagrammatically the nebulous spheroid in section with axis $a\,a'$ and equatorial plane $b\,b'$. The sun-forming central matter under condensation, at a certain period maintaining equal heat-radiants, would circumscribe S. The zone of planet-forming matter supposed to be moving at orbital velocity, upon condensation of the sun to a certain extent would have its annular centre of gravitation circumscribed about $p\,p'$. The lateral exterior matter $x, x, x, x,$ upon condensation would fall upon the sun or the planetary zone in gravitation-equation.

89. It will be seen that the zone-ring $p\,p'$, although detached, would still for a long time maintain its condensing condition, so that the central axis of the section of the ring might by condensation of surrounding matter become in time incandescent, or hotter than the nebulous sun. Therefore,

the sun in the interior of the equatorial orbit-plane of this zone $b\ b'$ would, through heat-exchanges, not suffer much loss of heat, whereas at the same time it would be radiating heat freely about its polar surface $a\ a'$ constantly into space to cause the sun's contraction in this polar direction. Further, after separation of a ring, during general contraction of the now separated central nebula its equatorial parts would keep up its extension in this direction by tangential momentum, and form a closed system by the attraction of matter towards the axis of the ring, which, as before stated, would become heated in proportion to the activity of its condensation.

90. Under the conditions proposed above, heat could be freely radiated from the polar regions into space to cause the contraction of the nebular sun in this direction; but this could not occur about the equatorial regions, where heat-exchanges with the condensing zone would prevent free radiation into space. The difference of local contraction-areas upon the surface of the solar nebula, in conjunction with the tangential momentum of its peripheral parts, would cause it to return to its oblate spheroidal form but of smaller volume. The spheroidal form would also influence gravitating matter to fall towards the plane of the solar equator, as suggested originally by Kant and shown more definitely by Newcomb for attraction toward an oblate spheroid*. All these conditions show that after separation of a zone-ring the system would again become oblate and in a condition to separate another planet-zone therefrom about its equator upon the same principles.

91. It will be seen that the surfaces of the sun and the planet-zone that would conserve their heat, therefore their nebular condition, would be those parts directly opposite to each other. Whereas the outer periphery of the detached zone-ring would be freely radiating its heat into open space,

---

* 'Popular Astronomy,' p. 513.

so that the contraction of the zone-ring upon itself by condensation would be *upon its outward and lateral parts only*, the inner parts retaining constant nebulosity by exchanges of heat with the nebulous sun's surface, as before stated. The contraction of the condensed outer matter would also be directed sunward by gravity, assuming the outer parts of the ring were moving at less than orbital velocity, which is necessary for the concentration of the system.

Accepting the theoretical matter for the case proposed above, if we take the centralizing globular nebular system of the sun at the formation of the Neptune-ring or planet-system, the diameter of the imaginary central globe at its full limits would then exceed the diameter of the orbit of Uranus; at the formation of Uranus it would exceed that of Saturn; at the formation of Saturn it would exceed that of Jupiter; and so on.

92. *Modes of Condensation of Interior Planets in a Spheroidal Nebular System.*—If we take the earliest conditions, the outer planets might very well go through the same transitions of the nebula as suggested above. We may observe, however, under the conditions proposed, that heat being always best maintained in the equatorial plane by exchanges with the newly formed planet-rings, where also the tangential impulse would be greatest, and the greater condensations going on constantly at right angles to this plane, where it was open to free radiation of heat, our assumed spheroid must constantly and rapidly flatten out its form of section in the direction of the orbit-plane of the zone-ring. Therefore the early planets, say Jupiter and those exterior to it, might be condensed from a spheroidal nebula under purely nebular conditions; whereas the inner planets assumed to be formed afterwards when the spheroidal solar nebula had become much flattened, and thereby presented much greater extent of radiation surface to depth of volume about the equatorial zone, may have sunk in temperature, so

as to produce exterior local condensation within the boundaries of the nebula. Under these conditions the continuity of the former nebular state would be wholly or partially changed. So that the planetary zone-system of Jupiter and its superior planets might somewhat resemble the condensation of our sun, but the inner planetary system might be wholly or partially formed from a condensation which took the primitive form of discrete or meteoric matter. This would partially account for differences of density, of rotation, and some other conditions which will be more fully considered hereafter upon discussion of certain exterior conditions.

93. *Mode of Condensation of the Extreme Outer Solar Nebula.*—In this we may possibly again find a modification or want of continuity of nebular conditions in relation to the planets Uranus and Neptune, which may be inferred from the direction of revolution of their satellites, under conditions already stated. In the case of these planets, from the want of concentrative force in the original nebula through weakness of effective gravitation due to distance, there would be a weaker centering-tendency of the peripheral matter. There would also be less heat radiated inversely proportional to the distance of the central heat-focus of the condensing sun. The tangential impulse being assumed sufficient to maintain the orbit of a zone-ring, local condensations might in this case occur at first to discrete matter, and the nebular system would thereby largely disappear. We may assume a gaseous system so extensive that the radiation capable of producing condensation would act superficially upon matter within a limited depth only of the nebula; so that gravitation would possess insufficient energy to draw this matter, if possessed of less than orbital tangential momentum in a gaseous state than that which would maintain it at its radial orbit, through the resistance of the interior nebula from the boundaries of a system so vast as that of Uranus or Neptune, under certain conditions suggested (§ 13) and others to be discussed. This

would produce a reverse direction of rotation, as will be shown hereafter.

94. *The Breaking-up of Gaseous Zone-Systems.* — The perfect state of equilibrium of nebular matter necessary for the complete formation of a planet-ring or zone (fig. 9, pp'), if such ever existed, except among satellites, could scarcely remain for a long time, as a slight disturbing cause at any position throughout the extensive orbit would destroy this equilibrium in such a manner as to permit the gravitation of its own mass to draw its parts together into the only form of static equilibrium that could be established, that is, a globe. Further, as before stated, in early planetary stages the intrusion into the solar nebula of exterior matter possessed of sufficient eccentricity to bring it at perihelion within the planet-ring orbit, would cause its inclusion within this ring owing to the resistance of the nebular matter of the ring itself to the continuity of its projection. In this manner all comets exterior to the system, or meteorites of great eccentricity, would be absorbed if brought in orbit-contact with the planet-ring. And although we may assume that the mass of a comet or of a shower of meteorites projected in a cometary orbit might not materially disturb the mean orbit of the ring-system, it might upon its intrusion possess quite sufficient momentum to destroy the equilibrium of a perfect ring, if such existed. This would not only be caused by its mass, but also by the local heat engendered, and the elastic expansion it would cause near the place of intrusion, together with the local drifting force due to difference of velocity and inclination of orbit between the orbits of the planet-ring and the intruded cometic or meteoric matter.

There is no doubt that if a planet-ring were perfectly symmetrical the inner attraction of its parts in a gaseous state might, under contraction through radiation from its own condensing matter, reduce its section until it might form even a liquid ring or rings; and this or these might again be detached into beaded strings of satellites, a condition which possibly holds in the case of Saturn's rings. A ring of perfect symmetrical condensation might finally part in one place only, and form a single satellite by its matter being drawn together by gravitation. This was possibly the case with our own moon, as will be discussed later on; but such perfect equilibrium of distribution of matter surrounding a gravitation centre could scarcely hold to the extent of a planetary orbit about the sun, and the zonal abandonment principle of Laplace is maintained if the zone is even imperfect in its circumference.

95. *Influencing Condition in Periods of Planet-formation. Critical Temperatures.*—Noting the irregularity of the masses of the planets, which cannot be accounted for by proportional condensations in time, there were no doubt present special inducing causes active for the time only, by which we may assume a greater or smaller nebulous zone-system was detached at any particular period from the central solar system. One of these causes was most probably the effect produced at certain times by the rapid condensation of nebulous matter to a liquid state at its critical temperature, within parts of the solar nebula, the mass of which we assume to be moving at less than orbital velocity at its periphery. Such condensed matter would, in the outer parts of the system, be immediately precipitated nearer towards the sun. This might not, as an early condition, be wholly possible with dissociated matter, that would only condense to a gas, which might suffer resistance from the elasticity of the highly heated interior. It might have occurred after the formation of the larger planets upon the uniform cooling of the entire system. If the temperature of the nebula was partially reduced at any period so as to cause it to pass from a gaseous to a vapourous state at any radius within the solar nebula, it is certain that the denser or metallic matter so

reduced to vapour would condense suddenly at its critical point, by a very slight further depression of temperature, to the liquid form. The general equable state of temperature of the nebula might permit, for instance, certain metals in the vapourous state to occupy large volumes, and to condense afterwards with a very slight depression of temperature at the critical point, causing a sudden interior collapse, and thereby the separation of a volume of peripheral matter which would afterwards maintain an orbit position consistent with its initial tangential impulse. These condensations might be at any distance from the sun within the nearly transparent nebula, according to the density and vapour temperature of the special element that was condensed. The amount of nebular matter, whether very voluminous or not, separated by the tension of an internal condensation and moving at about orbital rate at any time would after detachment maintain its free orbital position.

96. The interior of the zone-ring of detached matter would be the stress-plane of the exterior part of the critical condensation. Whether the matter condensed at its critical point remained as vapourous cloud in its precipitation towards or about the sun afterwards would depend upon the reaction of the heat of its condensation, radiated from the central system or sun at the time.

97. Upon the above-stated conditions, as far as they go, it is seen that the condensation of matter at the critical point would produce a permanent strain within the nebula, so that, seeing the nature of chemical elements and their very varied critical temperatures, the separation of planet-rings from the central solar system is not necessarily a uniform process depending upon a continuous condensation as proposed by Laplace. By condensation of interior matter at its critical point of temperature, a planet-forming zone or system may, upon this hypothesis, be separated from the sun at a certain time ; and for a long period after this separation the sun may

slowly condense nebulous matter upon itself only; until again, under certain conditions of critical temperature of the materials of the nebula, another planet-zone or system may be detached. Therefore, owing to the great differences in the critical temperatures of known matter, the zone or volume of detached matter may be of relatively large or small mass, and the planet formed therefrom will be consistent with this so far as the principle in question is active.

98. It will be observed that the condensation of any refractory metal from its vapourous state within the nebula would affect this particular metal only, and the vapours of other less refractory matter would remain in a nebulous state. Therefore, in any rotary nebular condensation, the denser, more refractory matter, moving at equal angular velocity, with the peripheral matter moving at nearly orbital velocity, must always drift to an inner position in spiral lines, being accelerated by gravitation also thereto. If the condensation remained, it would come to rest only as regards centralization when it attained an orbital position. Under these conditions internal planets must be formed of denser, more refractory matter than external ones.

A particular case of critical condensation would be one in which oxygen and hydrogen in a mixed state, below the temperature at which they must remain in contact permanent gases, were united into vapour. This condensation might be caused by a discharge of electricity, from an incidental chemical combination of prevalent elements within the nebular system.

99. The condensation to cloud, metallic or other, at the critical temperature of any annulus of nebula at a distance from the sun less than the radius of the condensing peripheral zone would obscure the exterior matter from radiation of the more highly heated central system, as the central heat and light would be reflected back from the condensed particles. This would cause the more rapid condensation of the detached

zone, whose tangential momentum would prevent its condensation upon the sun. This clouding effect would be reproduced at any following critical condensation of the matter of the sun system, and would again tend to condense a detached planet-zone or system; but it is not proposed that such a form of condensation is alone active in our planetary system. Other conditions have been already suggested, and will be further considered.

100. It may be noted under the conditions given above that a period when an element was under condensation at its critical point about the sun would be a period when his radiation would be materially obstructed. Therefore when an outer planet would receive much less of his heat. Such a period, which may in some cases have lasted many thousands of years for the complete condensation of a single element, may in recent geological time have produced a glacial period upon a wide extent of the earth, under certain conditions of distribution of land areas and direction of oceanic currents, which I have previously considered for geologically recent glacial epochs*.

101. *Modifying Conditions.*—In the construction given (§ 76) we assumed that an entire nebular ring, extending possibly within 10° from the sun on each side of the plane of the earth's orbit, was condensed to form the planet. This, however, we may presume was not the case. Very probably, as before suggested, the planet-forming ring was never perfect, or if perfect it is improbable that it should have condensed entirely at once into a single planet. Possibly condensations to meteorites form a common factor when the nebular system sinks below a certain critical temperature. Therefore, the earth's nebular ring might split up into a single planet and satellite upon one side of its orbit and be distributed in meteorites upon the other side. If these

* British Association Reports, 1885, p. 1020.

meteorites were of slightly different orbit-period from the earth they would finally unite with it at conjunction, but if of the same period they would maintain their *vis viva* and not be detected by any calculation in the variation of the earth's mean course or by telescopic observation. That is, assuming such meteorites to resemble those that fall upon the earth, which may have fallen from outer planet-rings, and which are generally of masses not exceeding a few hundred pounds.

102. The greatest condensation to form a planet would not in a uniform density-ring be at an intermediate position between an inner and outer planet, as may be inferred from the diagram fig. 9. The gaseous state would be best maintained towards the interior of the ring by heat exchanged with the nebulous sun. The condensation would, therefore, be on the outer surface of the ring at the greatest distance from the sun, where heat could be freely radiated into space, as before stated (§ 89). This exterior condensing matter, if it fell towards the sun, would cause the orbit-position of the new-forming planet to be towards the interior of the ring. In this position the precipitation of exterior matter falling in spiral path would give excess of velocity from gravitation to the interior matter beyond its original angular velocity, and might set the planet in gravitation equilibrium for a nearly circular orbit and in rotation at this inner position, although the original nebula had less angular velocity, as will be shown that it may have had further on. It is not necessary, therefore, upon the principle of exterior radiation to assume that a planet was formed entirely of matter of a nebulous planet-zone; it is much more probable that the inner condensations were at first upon the nebulous sun's surface, and did not separate therefrom until a dense motive system had been already formed in one position. Neither is it necessary to assume in all cases a single ring or a perfect ring-system; there may have been many imperfect or partial

DISCUSSION OF MECHANICAL PRINCIPLES. 79

rings detached before these formed a single planet, these being united afterwards by variation of time-orbits, and crossings of perihelia through eccentricity, or be drifting in spiral lines inwards.

The voluminous nebula of Jupiter would affect the nebular conditions in the formation of an inner planet. This will be best considered in relation to the formation of inferior planets, the Asteroids, and Mars, and the possible effects of this nebula upon the formation of the Earth. The subject will therefore be deferred.

CHAPTER VI.

Certain Conditions in the early Solar System which may be inferred from the Distances and Masses of the Planets upon the Nebular Theory.

103. *The Distances of the Planets from the Sun* appear to be in somewhat symmetrical order in individual distribution of position, although their masses do not indicate any law for their formation consistent with the condensation of a gaseous system or of a uniformly distributed discrete system by a decrease of density outward from the gravitation centre or sun. The approximately symmetrical order of distances, without relation to the amount of distribution of matter, was pointed out by Titius in 1772 *, which became known as Bode's Law owing to the special attention called to it by that astronomer †. It is illustrated in the following table, the scale of measurement being the sun to earth unit.

Table of Bode's Law.

	Mer.	Venus.	Earth.	Mars.	Aster.	Jup.	Sat.	Uran.	Neptune.
+	·4	·4	·4	·4	·4	·4	·4	·4	·4
B. eq..	·0	·3	·6	1·2	2·4	4·8	9·6	19·2	38·4
Theory.	·4	·7	1	1·6	2·8	5·2	10	19·6	38·8
Obs. ...	·39	·72	1	1·5	?	5·2	9·5	19·2	30·1

* See Miss Clerke's 'History of Astronomy in the 19th Century,' p. 87.
† W. T. Lynn, 'Observatory,' vol. xvi. (April, 1893) p. 178.

In the first line $+ \cdot 4$ is given as an arbitrary plus constant. It may be noticed that it is correct to observation, according to the law, for the places of the Earth and Jupiter, irregular with the inferior planets and Mars, and should be omitted altogether as a plus constant for the outer planets, failing entirely for Neptune. The second line in the table is in geometrical series from Venus, which is quite arbitrarily taken as 3. The third line gives the theoretical deduction of Bode's Law. The fourth line gives approximately the true distance as found by observation.

With Uranus and Neptune, there appears to be a certain element of orbital time relations; the year of Neptune being about double that of Uranus.

104. *The Masses of the Planets.*—No law has been discovered for comparison of the masses of the planets, except that the four inner planets are smaller and have a mean density more than five times greater than that of the four outer ones, which may indicate that their formation has been upon a different plan. In the outer planets there is a kind of proportion of masses to spaces, which agrees approximately with an assumed decrement of density of nebulous matter employed in their formation not inconsistent with the manner in which mixed gaseous matter would probably condense when placed around a gravitation centre.

105. Following the demonstrations of the theory of Laplace, the density of matter to form the planet may be estimated by reversing the process of its condensation; that is, by the dissipation of the mass of the planet into the assumed original nebular zone-ring volume it formerly occupied. Assuming that the planet will be formed upon the inner surface of the ring, as stated above, since radiation, and therefore contraction, must be exterior to this, we may for calculation take the mean distances from the sun of any pair of planets and make half their difference the radius of the ring—assumed for argument of circular section. Calling this r, the section of the ring

will be πr^2; making r_1 the mean radius of the orbit of the ring and therefore its circumference $2\pi r_1$, we have for the volume of the ring $2\pi^2 r^2 r_1$. We may conceive the ring of a certain oblateness; say this diminishes the area of the section by two thirds, our formula then becomes $\frac{2}{3}\pi^2 r^2 r_1$ for the corrected ring volume. Now taking the planet as a sphere $\frac{4}{3}\pi r_2^3$, r_2 being its radius, and dividing this into the volume of the ring, we obtain its assumed original density in the planet's specific density units. For comparison it is convenient to make the unit of density that of air at the earth's surface. Then the specific density of the earth is found by multiplying its volume by 5·6 its specific density compared with water and 800 the ratio of air to water. Calling this m, the complete formula becomes

$$\frac{\frac{2}{3}\pi^2 r^2 r_1}{\frac{4}{3}\pi r_2^3 m} = \frac{\pi r^2 r_1}{2 r_2^3 m}.$$

Other planets may be taken in a similar manner, changing m to m' according to the data for the density of the planet. The following table is taken from the above formula, adopting Bode's law for the mean place of the Asteroids:—

Table of Proportions of Densities of the Nebular Planet Ring-spaces to the Density of Air.

Mercury	1/1,050,000,000	Jupiter	1/505,900,000
Venus	1/81,230,000	Saturn	1/15,960,000,000
Earth	1/372,400,400	Uranus	1/248,400,000,000
Mars	1/418,800,000,000	Neptune .. (?)	1/5,000,000,000,000

106. The irregularity of these figures appears to indicate the improbability of the part of our nebula which formed the planets having been condensed from matter symmetrically distributed in an oblate spheroidal form, although perhaps this may not be altogether inconsistent as a form of condensation of the four outer planets and their satellites, as the table shows great increase of tenuity outwardly.

107. The decrease of the former nebular density-space between Jupiter and Saturn and Saturn and Uranus varies about as the powers of 1·17 to 1·12, and possibly if we knew its extent of original nebula, Neptune might follow a similar ratio. The figures appear to show upon the ring-nebula hypothesis that Uranus was condensed from matter weighing only about 4·6 grains to the square mile, a degree of tenuity difficult to imagine in a concrete system. This extreme tenuity might, however, be greatly reduced, if, instead of taking the oblate spheroid form before proposed we were to assume a more lenticular shape for our planet-forming nebula, thinning-out to a nearly flat plane, which would be quite consistent with observed forms of nebulæ seen in the heavens, of which we are supposed to observe the thinnest section *. Under this condition, the nebula might not vary much in density in the outer series of planets from Jupiter to Neptune inclusive.

108. Taking the oblate spheroidal form of nebula assumed for our solar nebula immediately before its condensation to planets, we may also suppose that effects of condensation through surface radiation produced some differences. The extensive and attenuated plane of Neptune and Uranus being open to this surface radiation would cause the nebula about the positions of these outer planets partially to condense into solid matter before the large masses of Saturn and Jupiter had considerably changed. Matter so formed and so widely distributed as it probably was originally before the formation of these outer planets, would scarcely condense entirely into a cohesive gaseous system, but we might more probably have at an early stage the formation of a motive discrete system composed of minute minor local condensation or of dust in the orbit-plane, according to the system of Kant and Faye. Such a system could only unite to form a planet if the original

---

* ♅ I. 200 Leonis Minoris, Plate II. *c*; ♅ I. 53 Pegasi, &c.

angular velocity of the particles were less than the tangential velocity of a particle in gravitation equilibrium according to the law of orbit, so that the particles separately condensed would fall in the direction of the nebulous sun of the period into elliptical orbits. Such particles would be of different orbit periods from differences of distance at the points of condensation, so that they might afterwards come into collision at the crossing of orbits about the coincident orbit position to cohere and form a planet; or they might unite in the exterior of the central nebulous system at the perihelia of their projection, which would henceforth become their orbit. This will be further discussed.

109. *Probable Form of the Original Planetary Nebula.*—If we take the original nebula to have been of about the same density at periods when the separate planets were abandoned, we may then plot a section that would represent the form which the planet-forming nebula would assume after a certain amount of condensation. This would show a considerable rounded projection over the positions of the planets Jupiter and Saturn, and as the Sun would also be contracting, this part of the nebula of the solar system would for a period assume a convoluted discoid form, possibly as represented in fig. 10; the positions of Saturn and Jupiter being shown at S, J. In the present construction we are considering the planetary forming nebula only which will be about the orbit-plane, and omitting all consideration of the circumscribing sun-forming pneuma system, which would maintain the spheroidal form.

110. In accepting this form, it is extremely probable that there was some external cause for the exceptionally large masses of matter in Jupiter and Saturn. Possibly these planets represent factors of an early intrusion of a local condensation of the nebula that formed at a distance from the solar plane, which was afterwards projected into it by attraction to the sun. Such projections might be constrained to

follow a circular orbit by the resistance they would encounter if they entered the sun's nebula at about the perihelia of their projections thereto. As before suggested, representation of a part of such a form of nebula seen in plan upon a large scale may possibly be found in the great nebula of Andromeda, Plate II., *a*, or in a more pronounced form in the ring-nebula of Lyra, Plate II., *g*. In fig. 10 matter is shown distributed

Fig. 10.

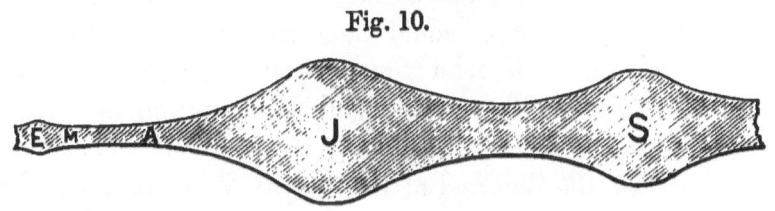

symmetrically about the orbit-plane; but it is more probable, as will be seen later on, that the planet-forming disc was not of this symmetrical form, but irregularly convoluted while still in a nebulous state.

111. *Effects of the voluminous Ring of Jupiter here proposed upon the intra-Jupiter Solar System.*—In assuming an extensive zone-ring of nebulous matter for the formation of Jupiter, the conditions that will be presented for interior planetary formation become materially modified from the uniformity of direction-gravitation due to the sun only, as before proposed. In the case before us, the intervening interior nebulous matter towards the Jupiter-ring would be solicited by two unequal gravitation systems, that of the sun and that of Jupiter; the Jupiter ring being practically active near its surface only. Assuming the intervening matter between this ring and the sun at the time to be wholly nebulous, this nebular condition being assumed to be largely maintained by the heat of the interior of the ring, there would then be a strong tendency through cohesion for the nebulous matter near this ring to be drawn either towards the sun or towards Jupiter in condensation. This would

occur particularly from the condensation due to radiation being greatest in relation to the depth of volume in the most attenuated part of the nebula, fig. 10, A. Therefore in this case the condensation would be from the surface in the orbit-plane of the nebula and would tend to break it up into separate superficial small local condensations. Further, from the great extension assumed for the nebular system of Jupiter exterior to the plane of orbit and the large volume of the nebulous sun at the time, there would be little excess of attraction towards the plane of orbit for nebulous matter within the orbit of Jupiter. Therefore, upon condensation of such a nebula, owing to the equilibrium of its position between the attractions of the Sun and of Jupiter, there would be a local tendency to form very small condensations or planets, particularly near the interior of the extensive nebular ring of Jupiter. These are probably invisible from our distance. Such small planets, at least the earlier ones formed upon the exterior radiating surface, would possess great inclination to the plane of the orbit of Jupiter in moving as free bodies under the stronger influence of the sun's attraction. In the equilibrium of position of condensation we have possibly the principal reason that the Asteroids are of small mass, and particularly that these small bodies should often be found moving in eccentric and inclined orbits, omitting extreme cases of inclinations probably due in part to other causes previously considered (§ 80).

112. *Relative Rate of Cooling of the Intra-Jupiter Sun-system.*—After the sun's spheroidal volume had retired from the inner surface of the ring of Jupiter's nebula, and this nebula was becoming of insufficient temperature to maintain an inner nebulous system, and many Asteroids of inclined plane of orbit had been condensed, the remaining nebula may have been disposed approximately as represented in fig. 10, where the letters S, J, A, M, E represent diagrammatically the positions of Saturn, Jupiter, the Asteroids, Mars, and the

Earth. Taking this outline of the section of the nebula and assuming surface radiation to be equal from all superficies, it will be seen that the narrow neck M A of the nebula of Mars and the Asteroids would by its shallow depth be the earlier part to cool down to condensation point, leaving Jupiter and Saturn still in a heated nebular condition. This condition of radiation remaining constant, the Asteroids, Mars, and our Earth would certainly be condensed to a liquid state long before the large mass of Jupiter could have lost its nebular condition. It therefore becomes probable that the condensation of Mars and of our Earth may have taken place much earlier or certainly not later than that of Saturn, so that our planet in a liquid or solid state would be much older than Jupiter. It is even probable that Jupiter may be considered only as an advanced minor solar system, and that at the present time it may not have its surface about its equator condensed to a solid coating : this will be discussed further on.

113. After the cooling of Mars and the Earth to a liquid or solid mass, the consecutive condensations of the inferior system of Venus and afterwards Mercury would again possibly more nearly fall into a system of periodic condensations similar to the earlier condensations of the outer planets, following the plan of consecutive exterior condensations as proposed by Laplace, except that with the inner planets we may have had a denser medium, or more probably a medium pervaded by discrete matter from which these planets were condensed, which would account for their superior densities and rotation periods—subjects that will be fully considered further on.

CHAPTER VII.

SUGGESTIONS FOR CAUSES OF DIRECTION OF ROTATION OF THE SUN AND PLANETS.—THE DIRECTION OF REVOLUTION.—VELOCITY OF ROTATION OF THE PLANETS AND OF THEIR SATELLITES.

114. *The Direction of the Solar Axis of Rotation.*—Let s' s'' (fig. 11) represent the nearest stars to our sun c, and the line of direction s' c s'' therefore the mean gravitation plane of the planets' orbits as before defined (§ 79). Let s, s, s, s be other stars more distant from the sun than s' s''. Bisect

Fig. 11.

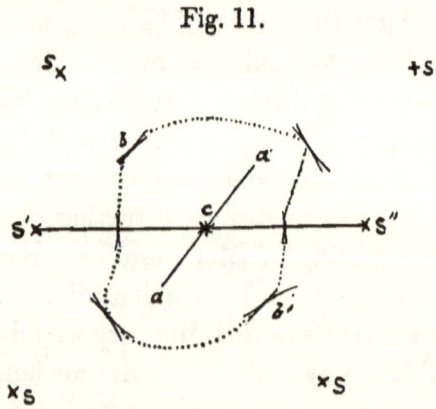

the distances between the sun and each of the stars, supposed of equal mass with the sun, by the arcs shown. The dotted lines embracing these bisections would represent the extent of the sun's original pneuma system, or that of the mean gravitation influences between the sun and these stars.

Assume the solar pneuma in revolution, when free from surrounding matter through condensation, with its axis at right angles to the plane $s'\,s''$. Then it is clear that a larger amount of matter in further condensation would fall to the sun in the directions $a\,a'$ than in the directions $b\,b'$; and as this matter would in condensation carry with it the original tangential momentum of the uniform angular velocity of the pneuma system, the equatorial plane would thereby be elevated from the original solar plane $s'\,s''$ towards $a\,a'$, the direction from which the greatest amount of gravitating matter would fall; so that the orbit-plane of planet-forming matter would be $s'\,s''$, but the momentum of the sun's condensation greatest in the plane $a\,a'$ and its equator be subject to the combined directive influences of $s\,s''$ and $a\,a'$. The plane of the stars taken above for illustration must be understood to be purely diagrammatical; no such plane exists, neither could the axis of original pneuma rotation be defined, the plane produced would be undulatory, but the principle will hold for stars in any direction from the sun and any plane of rotation. Nebulæ that approach the spheroidal form of which the great nebula in Andromeda may be taken as a type (Plate II., a) are generally asymmetrical, being more dense in certain directions *. Such systems in condensation would not therefore produce axes symmetrical in the plane of orbit.

115. *Direction of the Planets' Axes.*—If the suggested discoid form of the planet-forming nebula (fig. 10) existed with matter symmetrically distributed about the equatorial plane, the planetary axis would be vertical to the plane, but it is not at all necessary to assume the planet-forming matter to be placed symmetrically. If it were not so placed, the direction of rotation of a planet would be in equation with the mean momentum of matter falling to the planetary

---

* ♅ I. 200, ♅ I. 205, ♅ I. 53, M 81.

centre from any direction. Further, the pneuma system is assumed to extend in all directions and that condensations occur at its outer surface; therefore condensations would fall into the planetary plane in all directions, such matter being projected in cometary orbits. Matter thus projected towards the sun and towards the denser equatorial nebular plane would be absorbed in the nebular system and give local directive influences by causing intermotion within the nebula, without necessarily displacing greatly the positions of the planetary condensations then forming. This assumes the planet to be of much greater mass than the units of intruded matter falling constantly from various directions, and therefore such intrusions to be of insufficient momentum individually to materially disturb its orbital position. The projection of eccentric cometary matter would be very much more frequent in early times than at present, as all such projections would become absorbed in the solar-planetary nebular system.

Fig. 12.

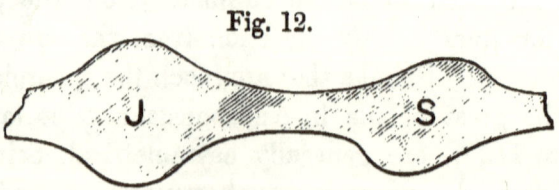

We must assume, from the extent of a pneuma-solar system, that cometary matter from exterior condensation would materially affect the intermotion of the parts of any system of nebula that was in a state suitable for planetary condensation, wherein every planet would form a gravitation centre with directive influences. Such intrusions only by directive impression according to the momentum of matter projected from any direction might disturb the mean plane of orbit of the planet, or induce obliquity of axis by diverting the mean revolution direction in the condensing nebula of which the planet was being formed. Under these conditions the theoretical general half section of the planet-forming nebula given in fig. 10 might be changed at the point of the commencement of planetary condensation to a section more nearly as represented by fig. 12 for the portions of the nebula about Jupiter and Saturn employed in their formation.

116. *Rotation of the Sun.*—As long as a spheroidal solar system of pneuma could remain in a state of gaseous cohesion entirely by the effects of its own elasticity through expansion by internal heat, so as to divert the direct action of free gravitation of its outer parts into a pressure upon the inner parts, such a system revolving upon its symmetrical axis in a frictionless medium or a vacuum would revolve entire. It would also revolve at equal angular velocity in all its parts as the least frictional form of motion for a continuous pneuma or highly gaseous system. In this case the peripheral velocity of the entire pneuma or its more condensed nebula must be such as will permit the exterior part of the nebulous matter to remain in contact or in cohesion upon the extreme outer surface of the entire system. This entails that the velocity of rotation of any part of the system must not exceed that which produces gravitation equilibrium according to the law of orbit. If the peripheral matter exceeded this orbital velocity at any time, it would be thrown off and depart from the central system. If the pneuma moved with a smaller velocity, it would press the gaseous matter towards the centre, forming a nebular system, and attain thereby a higher velocity by gravitation until the peripheral matter, maintaining its original angular velocity with the excess due to gravitation in falling towards the centre, had attained an orbital velocity, and then its pressure would cease.

117. Assuming such a nebular system as above defined, free from surrounding exterior attraction, moving at such a velocity that its peripheral parts could not rest in gravitation equilibrium as free particles, but must exert a pressure upon the system, then from any loss of heat or of internal elastic

force the tendency of all exterior matter would be active to press forward towards the centre, as the virtual velocity would be less in any interior part than that which would separately maintain its matter at the original distance of radius in a free state of orbital motion. Upon this condition the surrounding pressure within a gravitation system, if it exceeded the elastic force of the internal heat of the system for materials in a highly gaseous state, would form a dense mass or sun in the central part. In forming this central dense condensation, omitting the friction of the system which would produce heat, or assuming this equal to the acceleration of its gravity in falling sunward, we may take it that all matter condensed upon the sun would carry with it the linear velocity of its former position, which would be greater in the proportion of its original linear circumference to the circumference of the sun upon which it was afterwards condensed. This condensed matter would therefore rotate the sun in proportion to the mean excess of linear velocity or momentum of the matter condensed upon it from which it was formed.

118. To demonstrate the above proposition, we may assume that all nebulous matter formerly condensed upon the sun in consecutive shell-layers over its surface, at first from directive pressure of the interior parts of the nebula surrounding it, which we assume was moving in mass at equal angular velocity with the centre or sun. Then the condensations from the interior parts would impress small excess of linear velocity upon the sun in the early or central shells of condensation, and the more exterior parts consecutively higher velocity from the condensation of these more distant parts of the nebula in the outer shells, which theoretically would be consecutively brought down to the surface of the sun. So that with increasing volume the condensed central sun would attain constantly increasing velocity of rotation.

By condensation in the above paragraph, is intended purely

gaseous condensation,—the condensation being due to loss of heat by radiation from the system by which the whole mass maintained the elastic force with which it formerly resisted a central gravitation tendency.

119. We will assume as an hypothesis that our original solar nebula possessed the rotation period of Neptune of about 165 years. Then, assuming this nebula of spheroidal form every section of which was condensing by gravitation through loss of heat towards the sun, with directive momentum in proportion to the original virtual tangential velocity of its parts, the mass of the nebula being for the present problem taken to be in density inversely as the squares of the distance of its parts, then, if the nebula could be entirely condensed upon the sun to its present state without loss by friction more than the excess of momentum due to gravitation in falling sunwards, the sun should after such entire condensation possess a peripheral velocity equal to the peripheral velocity of the original nebula. That is, in the case we assume, the velocity of the planet Neptune in its orbit. This may be calculated.

120. The orbit of Neptune has a circumference of about 17,253,000,000 miles and his revolution period is about 60,000 days, equal to an absolute diurnal velocity of about 287,000 miles in our sidereal day. The sun has a periphery of about 2,679,400 miles. Therefore, if the whole nebula condensed upon the sun without friction in shell-layers within the orbit of Neptune, the sun's linear velocity would be at its equator equal to the velocity of Neptune in its orbit, which would make its period of rotation somewhat under $9\frac{1}{2}$ days. Observation shows the sun to rotate in about 25 days, so that the present diurnal velocity of the equatorial surface of the sun is 107,180 miles, that is only a little over 2·6 of that which would be due to condensation of the enclosed nebula within the orbit of Neptune, upon the conditions proposed above, taken as a trial hypothesis.

121. Upon the nebular theory the formation of any planet, say Neptune, could not have occurred until a large volume of exterior matter had condensed to this position. Indeed, the whole condensation must have taken place at the exterior or radiation surface of the nebula at a distance within which alone the planet could be formed, as exchanges of heat would prevent it condensing sunward, as before stated. Therefore the rarer portions of the nebula must have extended at an early period to a great distance beyond the orbit of Neptune, and the same extent of nebula must have been instrumental in sun-formation, although the condensation to form the sun might occur centrally from loss of general elasticity. If we assume the original radius of the nebula to be represented by $\frac{V}{v} = \frac{\sqrt{d}}{\sqrt{D}}$, where V is the orbital velocity of Neptune and v that of the periphery of the sun, and D the distance of Neptune, we have for the distance d of an exterior planet—in the present case for particles in gravitation-equilibrium according to the law of orbit—about 38,000 millions of miles, that is if the nebulous matter was moving with equal angular velocity to that of the sun's present equatorial surface.

122. This being the radius of equilibrium of a particle moving in a circular orbit at the distance given, would indicate upon the principles of gaseous condensation the fullest possible extent of our original nebula that could have been active in motive factors upon the sun, if its density diminished inversely as the square of its distance and the condensation was frictionless. In this theoretical calculation, therefore, we may take it that the angular motion of the nebula produced the angular motion of the sun, thus leaving direct gravitation in condensation to represent his heat.

123. In the above construction, although the motion of a planet may be used as the index of the extent of the original nebula at an early period, yet the planet's own formation may be excluded from the consideration of sun-formation, seeing

that the entire mass of the planetary matter may not exceed $\frac{1}{700}$ part of the mass of the sun and that the result of its formation may be entirely different,—the nebulous pressure on the sun by gravitation through loss of heat by radiation producing an intensely heated centre by concentration of exterior energy, as shown by Helmholtz (§ 10), whereas the energy of the planet system is expressed more particularly in motive factors. So that gravitation under certain conditions may increase the orbital velocity of a planet, whereas it may be active in heat factors on the sun.

124. *The Momentum of a Planet.*—If we assume a zone-ring of nebula formed of uniformly distributed matter or otherwise to be detached from the sun's nebula, and to be moving at the orbital velocity which would be necessary for its detachment, and that all parts of the zone-ring are moving *inter se* at equal angular velocity according to the theory of Laplace; then the entire momentum of the outer parts of the nebular zone, in falling towards the inner parts by condensation to form the planet, will carry with them the plus momentum of the outer parts which must appear in motive factors upon the planet when formed from condensation of such a system. If the angular velocity of the zone-ring at its outer parts were correlative with the orbital velocity of the inner parts, the gaseous system of the nebula would extend and not condense. It therefore becomes apparent, that to maintain the momeutum of the original angular velocity of a nebular zone, the zone must remain in some way attached to the solar nebula during a large part of its condensation, and this momentum must be conserved by the intermotion of its parts.

To consider the value of the motive factors of the parts of a condensing zone, we may take it from the period when its outer angular velocity equalled the orbital velocity of the next outer planet, that is at the period when its system was detached from the solar nebula. Upon this construction we

may divide the entire momentum of a nebular planet-zone moving under the sun's attraction into two factors of motion—angular motion and gravity. These we can distribute into two constants as the final condition of condensation:—
1. Permanent acceleration of rotation of a condensation within the zone, that is the future planet. 2. Acceleration of revolution to give the planet orbital motion. We may take it as an hypothesis that the sunward acceleration by gravity into the original momentum of the condensations of matter falling from the outer part of the system gives sufficient acceleration of revolution to establish the orbital motion: and that the difference between the angular velocities of the outer and inner parts of the zone-ring gives the rotational velocity to the planet as the probable action of the motive factors evident in the system proposed.

125. *The Orbital Velocity of a Planet derived from the centralizing gravity of the outer parts of a nebular zone falling upon the inner parts.*—Assume a planet newly forming by detachment of a zone at the periphery of the solar nebula, and that this zone is condensing most rapidly at the greatest distance from the sun. At this instant the periphery of the solar nebula must be moving at slightly less velocity than the detached zone, as before stated; so that we make the velocity of the outer periphery of the nebula of the next inner planet zone-ring that of the orbital velocity of the next outer planet. We have no data for this in the case of Neptune, which must be referred to the extent of the sun's nebula at its peripheral velocity, but we may take any other two nearer planets whose mass is sufficient to allow of their being regarded as nebular formations, say Saturn and Jupiter. In these the linear velocity of the sun's nebula at its equator could not have been so high as the velocity of Saturn at the time the Saturn zone-ring was just detached from the sun's nebula. If we consider these orbital velocities, we find Saturn equals 510,452 miles diurnal velocity, that of the next

inner planet Jupiter 689,855 miles; therefore, to define the velocity of Jupiter in its orbit upon the data suggested we require plus 177,403 miles. The plus velocity is herein assumed to be derived from gravity of the matter falling towards the sun, that is entirely condensing towards the position of Jupiter.

126. The simplest possible construction to show this is to assume that matter condensing from the outward part of the zone was moving just below the velocity of an inner part. Then this matter would not have sufficient momentum to maintain its tangential position, so that it must fall into an elliptical orbit of which its original position in the zone was its aphelion. If we assume that this matter could move without resistance, its velocity would increase according to the law of radii vectores until it reached its perihelion. And if we suppose this matter to form a planet at its perihelion position by encountering just sufficient resistance in surrounding matter to retain its perihelion-radius by deflecting it from an elliptic to a circular orbit, this would represent the velocity of the inner planet moving according to the law of orbit.

We cannot of course presume that this is the real condition, although it may represent its motive factors. The nebula being gaseous, would resist the direct continuity of the orbit of a condensed outer particle or aggregation of such particles, so that they could only fall sunward in spiral paths, possibly as flocculi; and the whole system might possibly be more nearly represented as a pressure system upon the new-forming planet accelerating its motion than as a free motive one, but the dynamic effects would be the same.

127. *The Rotation of Planets.*—Assume a planet ring to be detached from the sun's nebula, and that henceforth this is a free elastic body revolving with tangential velocity in equilibrium with the attraction of gravity upon it to maintain its orbital position according to Kepler's third law. If the

rotation of the peripheral band from any cause were slower than this, it could not leave the sun's surface. If it were faster its matter would fly off into space; or, putting the matter practically, the zone-ring would contract or expand to its position according to the law of orbit. As regards the separate parts of the planet ring, it is assumed that these being at first a part of the sun's nebula in a highly gaseous condition would revolve in all parts in relatively static positions, exactly as though the ring were a solid body attached to the solar nebula. This is inferred from the fact that any intermotion of its interior parts in a gaseous body would be more frictional than that of equal angular velocity in which there would be no internal displacements.

128. If therefore we assume that the planet ring revolved at first with equal angular velocity in all parts in mean gravitation-equilibrium for its distance from the sun centre, and that a condensation occurred in any part of the ring from pre-existence of a denser part which may have been caused by the intrusion of a comet, a shower of meteorites or otherwise,—then all parts of the ring-system sufficiently near together to support a system of cohesion in gravitation-continuity would be drawn towards the denser part with velocity of approach inversely proportional to the square of the distance from the gravitation-centre less the resistance by friction within the system. In this case the tangential velocity of the ring being assumed to be at its outer surface in equation with gravitation for a circular orbit, there would be no tendency for the ring to leave its orbit, but matter would be continuously drawn towards the centre of attraction of the new-forming planet, in which case accommodation must be found for the volume of matter set free from its outer zonal position and attracted towards the new-forming planet.

129. Under the conditions proposed, we should have towards the planet's nucleus currents from different parts of

the ring-system which would resemble in a certain manner similar currents in the atmosphere which find accommodation in cyclonic action in revolution about the centre of inertia of the motive system, as originally suggested by Descartes. In this manner the cohesion of the nebular ring would, in condensation under the circumscribing cyclonic action, tend to produce a rotatory nebular globe conserving a large part of the momentum of the original uniform angular motion it possessed in the former nebular ring when it was attached to the sun, and all the active gravitation forces brought about by its condensation towards the sun, deflected as they must necessarily be for accommodation of space within the new-forming planet's nebula.

130. If we assume that all parts of the planet ring conserved the momentum due to the equal angular velocity at the period the ring left the sun's nebula—which we are bound to do, or to account for the dissipation of this energy,—the rotation of the planet finally formed ought to be consistent with the linear velocities of its parts. Under these conditions the periphery of the planet when formed should rotate with linear velocity equal to the excess of linear velocity of the outer part of the nebular ring over that of its inner parts. This proposition may be discussed by the aid of a diagram.

131. Let $a\ a''\ a'''$ (fig. 13) be a part of the outer circumference of the nebulous ring, $b\ b''\ b'''$ the inner circumference towards the sun, and assume these circumferences to be bounding-planes of matter contracting towards the planet in the direction represented by a'' to a', b'' to b'. Let a'' to b'' be any small arc section of the nebular ring. Then the linear velocity of matter falling to a' being impressed upon a''—assuming a' to a'' moved with angular motion equal to that of the sun's centre, and a'' to be condensed upon a'—would set the planet in rotation with velocity proportional to the excess of the momentum of the matter condensed into the

inertia of the portion of the planet formed at the time. The condensation of matter from b'', when the linear velocity was less than the angular velocity of the planet at $a' b'$ in relation to the sun, would impress its momentum at b' in direction opposite to that of a', so that this would proportionally rotate the planet in the same direction as the outer matter from a''

Fig. 13.

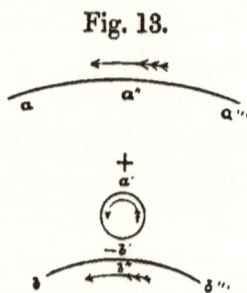

to a'. Now if all parts of the ring by accommodation conserved the momentum in condensation due to difference of length of arc and that due to attraction of the parts of the system upon themselves, by which the axis of the planet would be set in rotation by cyclonic action, the final velocity of rotation of the planet would be approximately such that its periphery would possess a velocity equal to the differences of the velocities of the inner and outer parts of the nebular ring $a\ a''\ a'''$ and $b\ b''\ b'''$, between which it was formed, upon the conditions of this proposition.

132. It is assumed that a planet upon final condensation from a nebular ring will be placed very near the interior of the ring, since radiation into space, as before stated, can only take place from the outer parts, and not towards the nebulous sun, which at an early stage of the planet's formation would be heated sufficiently to maintain a nebulous state and be nearly as large in diameter as the inner surface of the ring. The nebular matter being continuous in the ring, will maintain cohesion or separate by flocculation in attraction towards the planet, as before stated, during its formation;

so that the exterior condensation may be represented as showers of rain-drops falling upon the planet, carrying with them the tangential impulse due to their original motion and former position. This would occur so long as the interior heat of the system could maintain a purely nebular state. If the whole system was falling to the critical temperature of liquefaction of a large part of the nebula (§ 94), or the formation was influenced by a ring of exterior heated matter, as possibly was the case with planets interior to Jupiter, the regularity of condensation proposed above would be materially modified.

133. To maintain the condensation continuously in a nebulous state from the zone-ring to the perfect planet, we must evidently possess a large volume of nebulous matter in this ring, as any thinly distributed system would condense quickly by radiation of its heat after separation from the sun into discrete matter, before a nebulous planet could be formed. Therefore it becomes probable that only the large planets of our system, Jupiter and Saturn, could condense under purely nebular conditions wherein the equal angular velocities of the system might be maintained. Further inference that these planets are of purely nebulous formation is found in part from their low specific gravity, due probably to the conservation of heat in their interiors, resembling in a certain degree the conditions of condensation of the sun.

134. *Rotation of Jupiter and Saturn.*—Taking the planets Saturn and Jupiter under the purely nebulous conditions proposed, let r and r_1 be respectively the outer and inner radii or distances from the sun of two consecutive planets; then $r-r_1$ will be the diameter of the nebular ring which we may assume as a gravitation system of circular cross-section, as shown fig. 9, pp' (p. 67), from which the inner planet will be formed, and $2\pi(r-r_1)$ the difference between the circumferences of the outer and inner parts of the ring. Let P be the diameter of the planet, and πP therefore its circumference,

and H the number of sidereal hours in one complete revolution round the sun. The peripheral space that the equatorial surface of the planet will describe for its orbit in time will be $\frac{\pi PH}{2\pi(r-r_1)}$, and the theoretical term H_1, according to this proposition, will be $\frac{PH}{2(r-r_1)} = H_1$. The following table constructed upon these data will give the theoretical rotation-time in hours proposed approximately for a planet of equal density throughout formed from gaseous matter moving with equal angular velocity, upon this hypothesis. The T column is the observed time in hours as far as known.

135. It is seen that the virtual velocities of the parts of the condensation would depend upon their final position in relation to the centre of the planet. The nearer the centre the greater the angular motion impressed upon the planet from an equal linear velocity, so that a planet dense at its centre, and the reverse at its periphery, would upon final condensation have greater rotation-velocity than one due to the condensation of equally dense consecutive shell-layers over its surface to form its mass. In the following table the planets are taken as being of uniform density throughout.

Table of Theoretical Rotation of Planets formed under purely Nebular Conditions.

Planet.	Difference of circumference of ring $2\pi(r-r_1)$.	Circumference of planet $\times$ hours $= \pi PH$.	Theoretical diurnal rotation H_1.	Observed diurnal rotation T.
Jupiter ...	2,489,000,000	27,670,000,000	11·126	9·868
Saturn ...	5,534,000,000	56,896,000,000	10·27	10·16
Uranus ...	6,224,000,000	76,925,000,000	12·34	

136. In the above table the acceleration of rotation or of

orbit due to the gravitation of exterior matter falling to a nearer position to the sun is not calculated, as this may be taken up in acceleration of the planet's orbital motion, as before proposed. In the last column Jupiter appears of greater rotation-velocity than the theory demands unless it increases in density towards its centre; it would therefore appear, if the hypothesis given be strictly true as regards the only factor of motion, that the density of this planet must increase greatly from its centre to its measured perimeter. According to this hypothesis, to give a velocity of rotation of 9·868 hours for a planet of equal density throughout, its diameter should be only 75·248 miles. If it is partly solid and partly gaseous this number may express the diameter at mean density. It may be its diameter in four millions of years hence. Further the whole system of the planets may have been derived from nebular zones which at a period of early formation extended much beyond their present orbits, so that the planetary system may have contracted throughout its entire extent. This contraction would give excess of rotation beyond that resulting from the principles discussed above, even assuming a part of the energy may have been lost in the friction of planetary formation.

137. Although all difficulties with respect to acceleration of Jupiter's rotation upon the theory proposed may be removed by the suggestions given above, at the same time there is no doubt that the observations of the surface of this planet are very perplexing. The surface appears to be in violent agitation. It is not entirely improbable that this planet, quite recently as measured by astronomical past time, possessed a ring-system similar to that of Saturn. This ring, through some disturbing cause, the intrusion of a comet or other matter, was probably thrown out of gravitation-equilibrium, and the ring-matter, having stronger attraction to the body of the planet than to the parts of the ring so as to draw them together to form a satellite or satellites, the

separate parts of the ring are now drifting in concentric orbits upon and around the planet with greater velocity than that of its surface. In this case the heat engendered at the surface of the planet by condensations or collisions of meteoric matter projected thereon would produce a nebulous atmosphere permitting only the approach of the exterior meteoric-ring matter, which is probably in the form of dust, to drift under the resistance it encounters in spiral or cyclonic paths about the planet's equator and outward from it. This may produce the present surface appearance, which may be similar in motion to cyclonic areas in the atmosphere about our equator. The virtual velocity of the parts of the broken ring being greater than that of the surface of the planet, this matter in drifting over and covering the surface of the planet would present the only measurable part open to our observation, the planet itself being entirely obscured.

138. In Jupiter we have not the low density which we have in Saturn to indicate formation under purely nebular conditions, so that the reasons given in the two previous paragraphs are quite sufficient to account for the excess of velocity over that required by the theory of our table. If necessary, it would not be difficult to find others. If the zone-ring was originally elliptical, and the planet condensed at its perihelion position, the difference of linear ratios of the outer and inner surfaces (fig. 13, aa, bb) would fully account for the excess of rotation. Again, it is not impossible that exterior matter moving at greater velocity drifted into the planet during formation.

139. In the planet Saturn we have evidence of purely nebulous conditions in formation, in which the surrounding equilibrium of matter was so perfect that central rings of condensed free matter were possible of formation round its equator. In this case we have a rotation period in nearly exact equation with the tangential velocities of the part of the exterior nebula from which the planet was formed upon

the theory proposed. Nevertheless it must be in some degree accidental that this comes so nearly into agreement with my theory in this planet. It is in all probability much more dense in its central parts than at its periphery; it should possess, therefore, greater velocity than it does, but this difference may well represent the friction of the system in formation.

140. Further, with regard to Jupiter and Saturn, it is quite uncertain whether we make sufficient allowance for iridescence in measuring such bright bodies as these planets. The mean measurements of Venus as a bright body by Hartwig, Kaiser, Airy, and Ambronn give $17''\cdot593$ at a distance equal to its mean transit-position. The measurement of Venus by Auwers as a dark body in transit gives $16''\cdot801$, from which we may conclude the true measurement of Venus is probably $17''\cdot197$; applying a similar reduction to Jupiter, which it is impossible to measure in the same manner, we have its diameter about 80,000 miles. This proportion is also confirmed by some measurements made by myself of the iridescence of bright bodies under the microscope. It is evident also in the apparent thickness of the filament of an incandescent electric light, that may be compared with its reflection in a polished surface of black glass, which suggests a possible mode of measurement of Jupiter and Venus by reflection from a black surface.

141. Uranus is added to the table, which gives, upon the theory of the rotation of Jupiter and Saturn, a period of $12\cdot34$ hours. This planet, however, if we take the direction of its satellites as an index, is moving probably, but not necessarily, in the reverse direction, which might occur from its formation from discrete particles, according to the theory of Faye*. This theory may be shown by the same diagram. In this case matter is assumed to be moving in free orbits

---

* Faye, ' Sur l'Origine du Monde,' 2nd edit. p. 117.

round the sun, and the particles along the arc (fig. 13, $a\,a'\,a''$) assumed to have less virtual velocity than those along $b\,b'\,b''$, consequently the planet would move in the reverse direction. Another plan of rotation will, however, be suggested further on.

142. *Rotations of the Asteroids.*—The rotations of planets inferior to Jupiter would, of course, depend upon the mode of their formation. The attenuated plane of the nebula proposed for the formation of the Asteroids would leave these condensations about this plane so as to form an early break in the nebular system of the sun, and the henceforth separate nebular system of Jupiter ; so long as the Asteroids remained in the orbits of their original formation, they would take rotation periods consistent with the principles proposed for the formation of purely nebulous planets considered above, little affected by the decentralizing action of Jupiter or its original ring-system. Any of these minor planets, if formed through separate condensations of concretions of smaller planets, or of meteoric matter by after collisions, in the crossings of orbits, would have their original rotation due to condensation under nebular conditions diminished by the loss of the directive-momentum of rotation, which would be converted into heat at the time of collision. The absolute condition of these bodies must remain altogether speculative, our present limited knowledge being insufficient to obtain data for the actual rotation-periods necessary for its consideration. They most probably move with different rotation-velocities, and it is not improbable that some rotate in the reverse direction.

143. *Mars.*—If Mars was formed directly under the nebular conditions proposed for the superior planets, from a nebular ring which extended to the near asteroid Æthra, it would possess a rotation only slightly in excess of its revolution-period. If we assumed a wider area for the ring, possibly to include the formation of Æthra in another or opposite part

of the same zone-ring, we might theoretically give the rotation-period of Mars its actual or any value we please up to or beyond that of Jupiter. It is, however, probable that other than purely nebular conditions influenced the formation of this planet. Mars, the Earth, and the inferior planets possess a density-system which very probably indicates contacts or collisions of solid matter to form the interior parts. Therefore the nebular conditions proposed for the superior planets of low density do not hold entirely for index of rotation-period in this case for planets of high density, as it would be considerably modified by the discrete mode of formation.

144. Laplace suggested that a nebular zone-ring would break up into many parts, which would draw together and form separate nebular globes. These being formed at slightly varying distances from the sun, would finally coalesce and form a single nebular globe. In so attenuated a system as the zone of Mars would be upon this theory, such nebular globes as may have been produced by early condensations would probably be condensed by radiation, before union could be effected into a single globe. Therefore the separate globes or units of condensation might at a certain period be represented by a band of planetoids formed by condensations in different parts of the original nebulous band,—these planetoids being formed within the band at only slightly varying distances from the sun. If the orbits of the separate globes were elliptical and of nearly coincident period, they would constantly approach in opposition, come into collision, and by cementation, in which a certain amount of heat would be developed, produce finally after cooling the diversity of surface we observe upon this planet. It is possible that this form of collision may again occur at a definite calculable period with the near asteroid Æthra.

145. The rotation-period of Mars, as of other planets, must necessarily represent the sum of the directive momenta of

the factors of matter which formed the planet, and this might be fixed upon purely nebular conditions as with the planets Jupiter and Saturn, allowing time for the nebula to contract upon the sun only, without planet-formation as before proposed. The general conditions of density of the surface-formation of Mars and the presence of near Asteroids do not indicate equable nebular conditions at its formation. This does not, however, infer that it was not at one period partly a nebulous globe, the gravity of which was superior to that of any near body of condensed matter; indeed, this is probable for the formation of satellites upon principles to be discussed further on.

146. *Rotation of the Earth.*—If we take the same purely nebular conditions as proposed for the formation of the large planets by the condensation of matter carrying virtual velocities of the parts of the ring-system to the planet, and assume under like conditions that the earth's nebula extended at one period to the orbit of Mars, we shall arrive at a rotational velocity much greater than that which we observe. But in the discoidal nebular system proposed, fig. 10, p. 85, it would be extremely improbable that the actual earth-forming nebula in contact with it ever possessed this radius. By the principles of sun-condensation without planet-formation already discussed, and its condensation when matter just interior to its orbit was at a critical temperature, we may fix the earth's nebular ring-orbit consistent with its period of rotation, as before, by the formula $\frac{2(r_1-r)}{PH}$, considering the density of the earth and a certain extent of space between its orbit and that of Mars for its volume. The probability is that the greater part of the Earth-Mars interspace was condensed at an early period into separate smaller nebular planets, as just proposed for Mars, and showers of meteorites at its outer parts were moving in very elliptic orbits. The condensation which may represent the nucleus nebula of the

earth-moon was possibly at about its present mean radial distance from the sun. Under these conditions the outer condensations, whether nebulous, liquid, or solid, assumed of eccentric orbits, would be drawn at perihelion towards and into the earth's nebula, if these discrete condensations at the period extended, as was probably the case, beyond the radius of the moon's orbit. Further, in an attenuated ring-system, as the earth's system may have been at an early period, while rapidly cooling, condensations would occur at different parts of the ring, and at slightly varying distances from the sun, as before suggested, so that ultimate collision must occur between them and the earth. Under such conditions we should have through condensation variety of density and surface-conformity, and rotation in composition with all the motive factors of the earth's formation. The development of heat under the process of formation may have maintained a considerable nebula round the earth at an early period, but not being an entirely nebular formation its motion of rotation would be much slower than that calculated for Jupiter and Saturn if taken under purely nebular conditions. This matter will be reconsidered in detail further on.

147. The rotations of Venus and Mercury would be subject to the same conditions as that of the Earth and Mars. How far these planets may be formed from nebular or from discrete matter it is impossible to say. If formed from nebular matter, the motion would be greater than that of the Earth actually. If formed from discrete matter, produced by the general lowering of the temperature of the surrounding nebula to its critical temperature, it would be less; or these factors might act conjointly, so that the rotation might be *nil*.

CHAPTER VIII.

REVOLUTION OF SATELLITES.—DIRECT MOTION.—ROTATION OF THE MOON.—RETROGRADE MOTION.—COMPARISON OF THE REVOLUTION OF THE MOON WITH THE ROTATION OF THE EARTH FROM NEBULAR CONDITIONS.

148. *Revolution of Satellites with Direct Motion.*—In the revolution of satellites around their primaries under purely nebular conditions, no other law could hold for the condensation of nebular matter than that which must hold for condensation upon the primary; so that, if the motion of the planet's perimeter were equal to the difference between the linear velocities of the inner and outer parts of the primitive nebular ring from which it was condensed, the satellite's revolution should be consistent with this under the same mode of formation. Thus, taking the satellite's distance and revolution period, the virtual velocity of the satellite should equal the difference of rates of the extreme parts of the ring, previously expressed as $2\pi(r-r_1)$. At the same time this must be taken as representing the angular motion of the entire exterior nebular system within which the satellite or satellites, or any ring-system, as in the case of Saturn, were formed. As soon as the satellites were formed, the condition given, § 127, for planet formation must hold in the condensation of an attenuated system about the planet. The motion and position of the satellites, if there are more than one, must finally rest in relation to the planet according to Kepler's third law, the squares of the numbers representing their periodic times varying as the cubes of the numbers representing their mean distances, subject to such influences as the sun's attraction may produce upon the system. Under these conditions, the energy of the equal angular velocity of the original nebular ring may be split up into factors represented by the separate satellites and the general equality of the original angular velocity be lost, while retaining the mean momentum of the original surrounding nebular system about the planet. So that, in a purely nebular system not previously locally condensed, we ought to find the momentum of the angular velocity of the primitive nebula fairly represented in its entirety in the central system of the planet and all its satellites. Further, under these conditions of nebular or gaseous formation, the motion of the satellites must be direct with respect to the planet. Therefore it cannot include the systems of Uranus and Neptune, wherein the motions of the satellites are retrograde. This leaves us for consideration only the outer planets, herein presumed to be formed under purely nebular conditions, which have direct motion, or, more particularly, the satellites of Jupiter and Saturn. These satellites, as before stated, under the conditions given, cannot be taken individually, but as a mean of mass and motion of the whole contained in one system.

149. *Revolution of the Satellites of Jupiter and Saturn.*—Under the conditions given, let $2\pi(r-r_1)$ represent the mean momentum of the nebular ring active upon condensation in forming the planet and its satellites. Let

$$\frac{\pi S H}{H_1} = 2\pi(r-r_1) \quad \text{or} \quad SH - 2H_1(r-r_1) = 0;$$

where S represents the diameter of the orbit of the satellite, H the hours in the sidereal year of the planet, and H_1 the hours in one complete revolution of the satellite round its planet. In this manner, the revolution of the satellites of any system in the aggregate should agree with the angular rate of motion of the equator of the planet, and this with the momentum of the nebular atmosphere surrounding the planet from which the whole system of satellites is assumed to be formed. As experiment, we may take each of the satellites of Saturn and of Jupiter to be formed of equal quantities of nebular matter at about their present positions separately, and divide by their number so as to find the mean place of the imaginary satellite we desire to consider as the unit of the system. The result of this is shown in the following table:—

Table of mean Rotational Velocities of Satellites.

Mean position of	$\dfrac{\pi SH}{H_1}$.	$2\pi(r-r_1)$.	Difference.
Saturn's satellites and ring	5,264,300,000	5,540,100,000	− 275,800,000
Jupiter's satellites	2,777,450,000	2,490,950,000	+286,500,000

150. We see again in this an excess of velocity in the satellites of Jupiter which is consistent with the excess of motion over that derived from $2\pi(r-r_1)$ observed in the planet. This may infer an intrusion within the nebula previous to the formation of the planet of a large mass of matter moving with greater velocity, but so as to produce the same direction of rotation as that of the original nebular ring, as before proposed, which is again consistent with Jupiter's greater motion and abnormal mass. The difference shown, however, is made much less by taking the mean motion of the satellites in conjunction with their masses, the outer satellites of Jupiter being of ¿greater mass than the inner ones. And in the same manner the mass of Saturn's rings exceeds that of its satellites, so that this difference would be again diminished, and $\dfrac{\pi SH}{H_1}$ would come as fairly to my

theory as the conditions probably would admit of calculation in a system wherein there may have been intrusion of exterior matter, cometary or other, during the formation.

151. *Satellites of Mars*.—These small bodies move at a higher velocity than the equatorial surface of the planet. It is therefore clear that they could not have formed a part of a nebular system moving at equal angular velocity with the planet, assuming the planet was entirely condensed from the nebula and moved originally only at its present rotational period. By the condensation of a nebular zone as defined above for the satellites of Jupiter and Saturn, the rotation periods of the satellites of Mars would be made consistent by treating them as revolution systems, § 127. It is clear, however, as before suggested, that we have not in Mars a nebulous condensation of the kind that we have in the outer planets, where the condensation has produced a mass of small specific density. The probability is that the nebula of Mars possessed at a certain period a rotation consistent with the revolution of its satellites, but that the planet was of smaller mass, the whole system appearing as a planetary nebula. That about this period the planet entered into collision with one or more planetoids of smaller mass than itself, which were previously condensed to solid form. These planetoids may have penetrated its nebulous atmosphere without materially changing its rotational velocity, but have reduced that of the planet upon collision. Such collisions would develop great heat, partially liquefying the solid planetoid, and produce by cementation with it a partial protrusion of matter beyond the sphere, in which we have an index of the surface configuration. In this manner, the momentum of the outer parts of the nebula would be maintained, although the penetration by a planetoid to the centre would possibly cause sufficient disturbance of equilibrium in the nebula for the satellite-zone to immediately commence the formation of the satellite.

L

152. The orbit-position of a satellite can only follow Kepler's third law. In a satellite-zone open to a system of condensation at the outer radiation-surface of its nebula moving at equal angular velocity, in which refractory matter of the outer nebula condenses first, the condensed units will move through the resistance of residual gas under the influence of gravity in spiral lines, until they reach the central condensation; unless an orbital velocity position is found according to Newton's Law*; and in this position the satellite must be formed. Therefore a satellite may be formed at any distance from its planet. If matter falling upon a planet possesses less tangential momentum than that which produces orbital velocity at any position above the planet's surface, satellite formation would be impossible as such matter would fall to the body of the planet.

153. *The Moon* possesses a higher virtual velocity than the equatorial surface of the earth in the proportion of $2288 \cdot 43$, the moon's mean hourly motion, to $1037 \cdot 6$, the hourly motion of the earth's equator. Assuming the earth system to have been entirely nebulous, and the nebulous matter to condense in equal shell-layers over the earth, commencing from the time of the condensation of the moon from gaseous matter, decreasing in density inversely as the square of the distance from the earth's centre; then the earth should possess nearly the same initial velocity at its equator as that of the moon in its orbit, assuming no friction of formation developed into heat upon the earth's surface at the time. In many ways this is improbable, as the whole system does not confirm the earth's formation under entirely nebular conditions. It is therefore probable that the earth's nucleus was a considerable liquid, or partially solid, mass before the time of the moon's early condensation. Further, the configuration of the earth's surface indicates that it was probably formed in part by

---

* 'Principia,' Lib. ii., prop. xv.

concretion of planetoids or meteoric matter projected from exterior space into its nebular ring, which retarded its velocity and disturbed its axis of rotation. Either of the above-stated conditions, independently of the friction of formation or of tidal friction, if this may be included, would fully account for the earth's minus rotational velocity compared with the linear velocity of the moon, which must necessarily follow the law of orbit. At the same time it is highly probable that the nebular system of the earth extended much beyond the orbit of the moon at the time of the commencement of the moon's condensation *.

154. The rotation of the moon taking place in exactly the same time as that of its revolution round the earth, infers that the condensation of the moon was at first into a complete narrow ring, probably liquid at one period, therefore moving in all parts at equal angular velocity to the earth. If it condensed as a globe from a wide vapourous ring moving in all parts at equal angular velocity, as previously proposed for the outer planets, it would then possess a direct motion in rotation in excess of its revolution period. If it was formed from local condensation at first into discrete nebular matter, it would possess a reverse rotation †; combination of these factors, direct and reverse, might give its present rotation, but the proposition of the early formation being a narrow ring appears to me more probable to bring about the exactly equal periods of rotation and revolution.

155. The condition of condensation in rings which afterwards formed satellites appears to be difficult of conception to some astronomers. This must be so, if the density of all parts of the ring is assumed to be uniform, as it appears to be approximately in the rings of Saturn. But this system of equilibrium appears to be exceptional; if the rings move at

---

* See the Author's paper, Brit. Assoc. Reports, 1885, p. 915.
† See Faye's Theorem, 'Sur l'Origine du Monde,' p. 117.

orbital velocity, the equilibrium of its matter from tangential momentum and gravity is such that, irrespectively of its rotation, its matter may be considered to float on a frictionless plane ; so that the initial gravity of its mass acts directly upon itself. Therefore, if the nebular moon-zone were unequally distributed, its condensation would be direct to the densest part.

Suppose the condensed zone represented by fig. 14 a, then its point of tension would be at x. Assume this point to separate by the initial slow action of the entire gravity of the ring. Then its matter near the point of separation would gather into two semi-globular terminals, fig. 14 b, x'. Initial

Fig. 14.

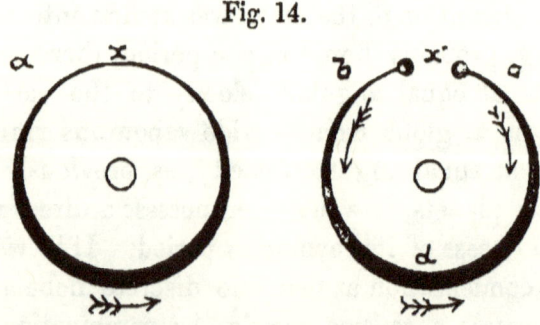

gravity would now act in opposition to the direct momentum of the mass, accelerating the velocity of the limb b and retarding that of c. So that at some point about d the ring would condense into a globular mass, by its extreme condensations coming together.

156. *Retrograde Motion of Satellites.*—The conditions proposed for the direct motion of the satellites previously considered as derived from the condensation of the gaseous nebula, could not possibly hold upon the hypothesis given for the satellites of Uranus and Neptune, which move in the retrograde direction. It may appear, so far as the direction only of this motion is concerned, that this would be demonstrated by the theory of M. Faye, § 141, as due to their

formation from discrete matter, the satellites being condensed from a ring of such matter, every particle of which was originally moving in a free orbit in gravitation equilibrium; this theory may possibly be applicable to the satellites of Neptune. There are, however, many peculiarities about the satellites of Uranus which do not admit of this hypothesis. They move in orbits inclined nearly 80° to the planet's orbit-plane, so that the differences of velocity of the parts of the orbit of any assumed nebular ring or matter caused by the differences of distance of its parts from the sun would be very small. These conditions would, therefore, cause the satellites to revolve at a very slow rate, that is, at about $\frac{1}{200}$ of that observed. We must, therefore, certainly look for additional causes for which other suggestions may be offered.

157. If discrete matter was formed at the limits of the solar nebula, when this was moving at less than orbital velocity, as before supposed, and the early discrete matter was formed of the more refractory matter so as to leave the more attenuated, less refractory nebula to form a resistance to the centralizing condensation of the discrete matter, then this matter might carry the momentum of its former angular velocity to the inner, denser nebular matter which formed the sun at the time, leaving the residual matter slowly condensing at less angular velocity. Such a system would produce a solar nebula moving at higher velocity than its peripheral lighter outward parts. I have endeavoured to show in my work upon Fluids* that every fluid system in rotation, cyclonic or other, engenders in the surrounding fluid or medium which offers resistance to its direct motion an opposite direction of motion of rotation around its borders. By this action the central rotating fluid attains a kind of rolling contact upon the

---

* 'Experimental Researches into the Properties and Motions of Fluids,' 1881, p. 224 *et seq.*

surrounding, more static fluid, which produces by this mode of rotation the least frictional resistance to the motion, in what I term friction-whirls to the direct flowing system. Upon this principle, if the central system in this case were condensing the more refractory pneuma into nebula with increase of angular velocity by the effects of gravity, then the surrounding, less refractory pneuma, less motive in the direction of the central system by the resistance of surrounding matter not moving at equal velocity or possibly in the same direction, would offer a certain amount of resistance at the periphery of the central condensation of nebulous matter. This is shown by the lateral form of motion to a current, fig. 5, p. 44; it may possibly be better explained by a diagram.

Let one plane of the interstellar space, subject to the superior

Fig. 15.

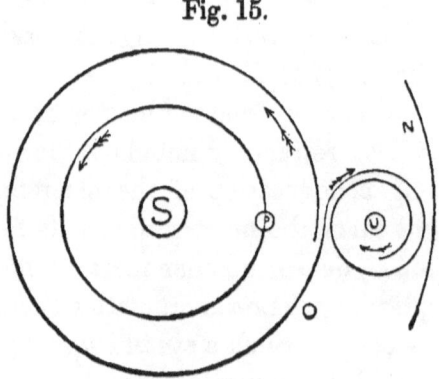

attraction of our sun, be represented by the circumscribing outline N, fig. 15, along which line the sun's attraction is assumed to be in equation with that of other near stars. Let S be the centre or sun; and let the central nebula be bounded by the circle O, the arrow near the line showing the direction of rotation. Then in the space bounded by N on one side and O upon the other the matter would be less motive than that of the central system; and this would rotate U in the reverse direction, as shown by the arrow. This form of motion is shown by many experiments in my work on "Fluids" already quoted.

158. If the central system contracted by the influence of gravitation and other conditions already discussed, say, to the diameter represented by p, which we may take at the orbit of Neptune, then the whirl system U would be also contracting upon itself; and being also attracted by gravitation, it would become a motive part of the solar system S, falling into the same direction of revolution but with the opposite direction of initial rotation, just as a roller or ball-bearing moving between the surfaces of an inner and outer cylinder progresses along the surface of the inner cylinder in its direction of revolution, but at the same time takes the reverse direction of rotation. Upon this theory it may be inferred that Neptune and Uranus were possibly perfect rotatory systems of nebulous matter before they were incorporated into the central or solar nebular system.

159. In the above construction, considering the inclinations of the orbits of the satellites of Uranus, we have at first the loss of directive momentum due to the differences of velocity of the parts of any ring-system that could have formed and rotated these satellites in the plane of the planet's orbit, so that the surrounding matter in the satellite's orbit-plane became more subject to resistance than if the planes of orbits of the planet and satellite were nearly parallel. The motion for reverse rotation under fluid resistance is active laterally, exactly as in the plane of motion, and resistances are more likely to occur exterior to the orbit-plane than in it. The form of nebula at a certain period of formation best adapted to produce the system of the satellites of Uranus, is possibly that of ♅ I. 176 in Coma Berenices, where the nebula appears to turn up at one terminal at a considerable angle to its mean plane of rotation.

160. Under the same principles as given in the above

hypothesis, it may be suggested that the contact form of reverse rotation may have been produced in a nebular condensation having either no rotation or a reverse rotation, being attracted into the sun's nebula. Such a condensation may have occurred in an interspatial position between the nearly equal attractions of several stars, as, for instance, in the position outwards between b and a' or a and b' of fig. 11, p. 88. If a nebula so formed drifted by small excess of gravity towards our sun when it was in a nebular state, it would enter the frictional system of the borders of the attenuated solar nebula. The inertia of the newly introduced planetary nebula would resist the rotation of the solar nebula, which would therefore produce a motion of rolling contact in the meeting-plane between the two systems, as before suggested, as the least frictional form of fluid motion. In this manner a kind of wheel-and-pinion motion would be induced, in which the smaller planetary nebula would represent the pinion moving in a reverse direction to the solar nebula wheel.

CHAPTER IX.

COMETS CONSIDERED AS ORDINARY GRAVITATIVE MATTER IN ROTATION CONSTRUCTIVELY AS A PART OF THE PLANETARY SYSTEM.

161. If we can accept the idea of a universal pneuma, § 23, and that this pneuma was motive in rotation, separating and condensing into separate systems, § 43, a condition of original motion in matter that may not be unlike that proposed in the theory of Descartes and Faye [*]; then all separate parts of the system must continue motive under condensation to take up the general momentum of the system.

162. If we consider the relatively small volume that would be circumscribed by our solar nebular system, if taken to be in its original nebular state of spheroidal form at a period when it was circumscribed within the orbit of Neptune, as compared with the mean distance between our sun and the nearer surrounding stars, we find what a relatively small space the orbits of our planets occupy. So that if the original pneuma at its earliest period extended to the mean of interstellar space about our sun as suggested, we can only imagine that many millions of local condensations due to exterior radiation were formed in exterior parts of the system. These condensations would afterwards only slowly drift sunward, but they would still carry with them the same factors of original rotative influences and the influence of the attraction to near-surrounding matter, as before discussed, § 70.

163. Therefore, taking the original pneuma system as shown

---

[*] 'Sur l'Origine du Monde,' 2nd edit. p. 101 *et seq.*

in one imaginary plane, fig. 8, A B (p. 59), extending to the mean distance of our sun and a near star, we can but imagine that at an early period there may have been many millions of local rotatory systems of matter condensed to a nebular condition in a free state, which would be moving from the outer pneuma system from the positions a, b, c, d sunward by central attraction. We may group such systems together as comets.

164. In the above construction we may suppose that the cometary system was the earliest prevailing system of the outer condensation of matter; that the comets which were formed by local condensation to nebulæ were most generally absorbed into the solar-planetary system when this was of immense volume; so that at a certain period our solar system would have appeared, if viewed from a great distance, as an immense floccular system wherein the exterior cometary condensation would appear incandescent from friction and electrical excitation within the interior nebular condensation, and from intense chemical action in the condensation of the pneuma to nebula. These flocculi would be drifting sunward by the effects of their attraction and the small resistance of the surrounding attenuated pneuma in spiral paths, similar to the spiral nebulæ 31 M, 81 M, 56 ♅, 168 ♅, &c., as before proposed.

Another condition that must mark the formation of comets considered as exterior condensations of the pneuma system, is that the direction of orbits must have been influenced by the gravitational effects of the larger mass condensation forming at the time upon the scheme proposed in § 83 and illustrated by fig. 8, p. 59. Therefore, comets must have been formed in many cases in series, from attractions a, b, c, d of this figure, taking one orbit-plane, but with varying eccentricities of orbit, depending upon the amount of original tangential impulse each comet possessed upon starting sunward, as the conditions entail.

165. Upon these principles the comets and cometary matter

we possess at the present time in our ancient solar system represent only the waifs and strays formed by exterior condensations which have escaped absorption, that have arrived from interstellar space after the sun had condensed to nearly its present volume, § 68. The relatively small number that remain may also, through the disturbing influences of the antagonism of initial centralizations and solar attractions, hereafter experience internal strains and collisions within their systems about perihelion which may cause their disintegration or deformation, so that they may be lost to astronomical observation in the future, under conditions to be discussed.

166. Although the earlier condensations of pneuma to nebula exterior to our motive solar-planetary system would produce, upon conditions discussed, rotatory systems of nebulæ, we can scarcely imagine that such nebular conditions could be generally maintained in exterior matter until the present time under the excessive radiation of their original heat into space. Therefore, we have the extreme probability that cometary matter is at the present time largely condensed into solid small units of meteoric or planetary matter.

167. The theory of association of comets with meteorites is as old as Anaxagoras, who states that comets are the congregation of wandering stars that approach so near to each other that they appear to touch*. This theory, for certain factors of cometary existence, is greatly strengthened by the discovery that our known meteor-streams follow in the orbit of certain comets, as pointed out by Schiaparelli for the August meteors, and for other meteor-streams by Oppolzer, C. F. W. Peters, Prof. Newton, and others, which theory is also ably supported by Prof. Lockyer †.

Under this theory the difficulty of finding cause for the self-illumination of such streams, if they form comets, is very

---

* Stanley's ' History of Philosophy,' p. 64.
† Astr. Nach. No. 1384. Lockyer's ' Meteoritic Hypothesis,' p. 138.

great. Professor Tait has endeavoured to prove that such meteorites would be subject to sufficient collisions among themselves to account for the light *; but if the meteorites follow one another in streams, there must be very small differences of velocity between them, and if they approach one another, assuming them widely distributed miles apart, their collisions must be very gentle, and produce very little light, if any. Otherwise, the observations of Mr. Denning upon the Biela meteors show that they possess great diffuseness of radiation, so that their paths appear to diverge from an area rather than from a point of the sky, indicating intermotion among their separate units.

168. There is another objection to the meteoric-swarm theory which has not been attempted to be met, or even suggested that I am aware of, which is, that the separate units of the comet must, under the conditions of this theory, necessarily follow Kepler's third law of orbital motion, that the squares of the members representing the periodic times of the separate meteorites must vary as the cubes of their mean distances from the sun. Therefore, assuming any comet to have a tail of one million miles in diameter, as commonly observed for the larger comets, the velocity of the outer meteorites of the swarm must be very much less than that of the inner ones nearer the sun. Under this condition, if the comet were a swarm and at a certain period from some unknown cause of the symmetrical form common to comets, as, for instance, that of Halley or of the great comet of Sept. 1882, or other, Plate III. h, i, j, as the parts of the swarm would be actuated by various velocities, it could retain this form for a short time only. So that on its return to perihelion, for instance, it could only be represented at most by a scattered band of meteorites spread over a great distance in space that could never again appear as a comet. To take a self-evident

---

* Edin. R. Soc. Proc. 1879, p. 367.

case of the conditions in question, assume our moon at opposition to take a solar orbital motion without revolution around the earth, then it is clear by Kepler's third law that we should soon leave it behind us in space, so that it would become in time in conjunction with the opposite part of our orbit. Indeed the only condition possible for a meteoric theory in which a comet can retain a symmetrical form, is that the system of meteorites that form the comet should be in revolution about the centre of inertia of the system, moving in elliptical orbits, in the same manner as the comet moves about the sun and as satellites are in revolution about their planets. This will be more particularly considered presently.

169. *Comets of long period.*—The general principles of direction of orbit from local condensations at a great distance from the sun have been discussed, § 71. We have now, therefore, only to consider the probable intermotion of the parts of such distant condensations as may possibly produce comets of the symmetrical form we observe in them upon planetary conditions, and, therefore, such as are outwardly, as it appears to me, evidently moving under the direction of symmetrical orbital law.

In the extensive volume of pneuma considered as the extreme field of comet-formation, which would be subject to the influence of the near stars almost as much as that of our sun, § 76, the centralizing influence of gravitation would have little effect in changing the natural formation of individual systems of matter after they were once constituted. Therefore, assuming original motion in the pneuma such as we have found necessary for the formation of our solar-planetary system, § 62, such motion must, as before stated, have extended to all parts of the solar pneuma. If any original isolated system, of large volume in its original state, were in slow revolution with its parts moving at equal angular velocity, then upon its condensation to a smaller volume its rotative velocity would increase, as previously discussed for solar rotation, § 116. This rotation of any part of the system would be maintained in projection sunward, and if in free matter projected from a great distance, it would form a comet of long period.

170. *Comets of short period.*—These possibly depended in many instances for their orbit upon deflection of the matter of the comets of long period by the influence of planetary attractions. At any early epoch such long-period comets, moving at high velocities by accumulated gravitation in falling from a distant part of space, would fall into the solar-planetary nebula or into any detached zone-ring of nebular matter which may have been present at the time moving at orbital velocity round the sun. In such a case the motion of the comet would be retarded by the nebulous matter and enter into composition with the motion of part of the zone, or, if not incorporated with it, it would be deflected by it from its original orbit into a less eccentric orbit. Comets of short period would also be formed upon local disturbance at any part of a planet-forming zone-system by a local condensation forming at a distance within the orbit-zone from the position of the planet's condensation, which was at the time beyond that of its prevailing attraction. Comets so formed may be termed *planetary comets*, and bear relation to certain planets, as Saturn, Jupiter, Mars. Short-period comets were also probably formed by local condensations outside the mean planetary plane at the same period as the planets were formed, when the planet's mass was not the superior attraction with respect to the position of condensation. They may also be formed by the detachment of parts of the tails of long-period comets disturbed at perihelion by disruption of the cometary matter by heat, which retarded the revolution-velocity of a part of the system.

171. *Symmetrical elements of Comet-formation.* — I have offered some general arguments for the construction and motions of local rotative systems formed in space and pro-

jected towards the sun by its superior attraction, as suggested above, in a paper published in the 'English Mechanic' in 1883 *. These ideas will be now reproduced, with some extenuations that appear to me necessary in reconsidering the subject.

We can scarcely enter into the discussion of the motions of comets without making a clear division of the subject as to whether they have been derived from purely nebular conditions directly, or have suffered from the attraction of other bodies near which they have passed so closely as to have their general motive symmetry destroyed. That this condition of disturbance occurs is quite evident in the present state of what was formerly Biela's comet, and in some comets which have from unknown causes become invisible, and may now only be represented by wandering meteoric systems, but whose orbits were clearly defined upon their first appearance. It follows, therefore, that for any symmetrical law of comet construction we must take into consideration such comets as retain the symmetrical form which we may suppose that they possessed when originally projected from space. These comets may be the only ones that are not under conditions of dissolution, which may be at the present time the more common phase of comet life. If the comet depart through some disturbance from its law of original construction, its matter may present afterwards only what we may term a specialized confusion, too complicated to discuss by the most advanced science.

172. The symmetrical comets will be presumed to possess elliptical orbits which have not been materially disturbed from the time of their original formation and projection towards the sun. Types of such comets may be found in Halley's, Donati's, the great comet of Sept. 1882, and many others of long period. They may be distinguished by possessing

* 'English Mechanic,' 22nd June, 1883.

a head and forward projection of the coma of symmetrical outline, after the manner of h, i, j, Plate III.

173. It has been fully demonstrated that the head of a comet follows a truly elliptic or parabolic orbit; so that we have no doubt that this part of the comet is subject to purely gravitational influences. The difficulty presented by these bodies is that the tail has not conformed to the conditions of orbit for the parts of a free system, or as it should do upon the swarm theory. This point will now be particularly discussed.

174. *Comets considered as Gravitative Matter.*—The certainty that the orbits of comets conform to the laws of gravitation was clearly laid down by Newton as a principle. This would lead us to infer that they are composed of quite ordinary gravitative matter, which is again to a certain extent confirmed by the spectroscope. The reason why it is thought that there must be a deviation from this law (by Olbers, Bessel, J. Herschel, and others who have followed this idea) is that during the perihelion passage of the comet, the tail, which must be considered a very material part of the cometary mass, diverges greatly from the normal elliptic or parabolic orbit.

175. To meet this case, we have been asked to assume that the tail is unlike any form of matter with which we are acquainted, that it must be *antigravitative*, or that it becomes so from some cause at or near the perihelion passage of the comet. It may be suggested that the law of universal gravitation is one of the last we should abandon, seeing that it has done such perfect service wherever our knowledge of the conditions was exact. Further, there are other sufficient reasons by which we may conclude that there cannot be repulsion in any part of the comet, as the centre of gravity of the cometary mass follows constantly in the true orbit. For if a portion of the cometary mass, that is, the tail, *changed its state of constant attraction so as to become unlike other,*

COMETS CONSIDERED AS GRAVITATIVE MATTER. 129

ordinary gravitative matter, and this portion became *repulsive* from the sun by heat, electricity, or otherwise, when the comet passed near perihelion, then the centre of gravity of the cometary mass must be altered by this repulsion, and *would be displaced in relation to the sun's attraction; so that the sun would no longer occupy the focal point in the orbit of the portion of the cometary mass that remained attracted to it.* Therefore, a new form of orbit must necessarily be formed to suit the altered conditions of attraction, gravitatively central to the mass attracted only, or there must be in this case a deviation from Kepler's third law, a condition at least improbable. As regards the change of form or properties of any known matter, so far as physical knowledge extends, we may take it that heat has no effect whatever upon its ponderability, so that difference of gravitation from this cause would not be possible through the difference of temperature induced in the comet by passing very near to the sun. Therefore, although the gaseous matter might be expanded, its centre of gravity would still follow as nearly as possible in the cometary orbit, not *diverge*. In other words, from the fact that no considerable change of orbit is evident *after* perihelion passage of a comet, we may conclude that the centre of gravity of the cometary mass traverses the true orbit of its projection in space, carrying with it the mass of the tail, and thereby that it conforms to planetary laws in the same manner as any other entirely gravitative system of matter.

176. At the time Sir John Herschel suggested that the tails of comets might be of a kind of matter of which we have no knowledge, which is antigravitative in relation to the sun, the spectroscope had not been brought to bear upon cometary matter. We now know that carbon, hydrogen, and other elements form constituents; so that the introduction of imponderable matter can scarcely be permitted, even hypothetically, in this case. Further, it has never been explained in this theory, how the cometary matter, after it is expelled

K

from the sun, recovers its attraction or cohesion, so as to re-form the actual comet after perihelion as we know from actual observation that it does. If we adopt the theory of electrical repulsion as proposed originally by Olbers, and supported by Bessel, Norton, Zöllner, Bredichin, and others, which is now most popular, this in no way relieves the difficulty. If the tail is repelled on approach of the comet to perihelion, with an internal separative force due to electricity of one sign assumed to exceed gravity, it must necessarily be left behind, and can never regain the orbit velocity of the head of the comet. Now this is precisely the opposite of what is requisite to represent the motion of an actual comet; what is required is that the comet shall be elongated at perihelion, for which the action of gravitation alone is sufficient, and that the matter of the tail shall have its velocity of direct projection *increased* in such a manner that it shall describe larger arcs at the radii of its separate parts from the head of the comet, to which the head remains constantly as a centre during its perihelion passage. It has been suggested that the tail forms a small part of the cometary mass and may be re-formed from the matter of the head : this is in the highest degree improbable, as it is not necessary that a comet should even possess any head—for that which represents the head is often merely the centre of inertia of the cometary system which follows in its orbit.

177. *Conditions under which a Comet may be considered as a Planetary Body.*—Taking the evidence of apparent conditions of comets generally, we find them immense volumes of what appears to be nebulous matter of somewhat symmetrical outward form. Therefore, evidently forming *systems of matter* held together by internal forces which must in some way conform to the laws of gravitation, and so far resemble planets. We find comets otherwise of very small density, as is evident from their not disturbing the orbit of a planet whilst passing near it. Therefore, to account for such

immense volume and small density, in a solitary or planetary-like system, we have in the first place to consider the possibilities of ordinary gravitative matter, which we now know it to be, being held together symmetrically by a system of forces, wherein the matter itself, although this is in a state of very great tenuity, remains practically an adhesive system.

178. Now, following the analogy of things known to account for an enormous diffusion of matter from or about a central attraction of gravitation, being either engendered or maintained in an extensive nebula or planetary-like system, such as a comet may be considered to be; we have only three known conditions which may so far react upon a gravitating or *centralizing* force in matter to insure this state of diffusion, tenuity, or decentralization:

1. *Heat-forces may separate the parts of a unit system of matter to any degree of tenuity.* 2. *Electricity of one sign may separate attenuated matter similarly to heat but with greater activity.* 3. *The tangential action of the revolution of the outward parts of a gravitative system about a centre or focus may separate these parts proportionally to their velocities and distances from the centre according to the law of orbit to any degree of tenuity.*

If we consider the probability of the one or other of these forces being entirely or principally active in a cometary system, we find, with regard to the first, that to maintain a degree of heat sufficient to diffuse gravitative matter in a nebular form to such extreme tenuity as we witness in comets, we must assume great intensity of this heat, even if we assume the central attraction small. Further, for this heat to act as a *separative* force, we must assume a permanent *gaseous state*, as heated solids could not repel one another to produce the observed volume. Then, again, if we assume the heat present to be sufficient to account for the extreme diffusion necessary to produce the known tenuity, still we have the difficulty present that this heat will be subject to constant radiation in space, from all exterior parts of the system, and therefore the comet be subject to a constant loss of the *decentralizing* force, which alone in this case could support its tenuity. As we know that the larger comets pass to very distant regions, where little heat can be derived from the sun, we can scarcely imagine that heat sufficient to maintain the enormous diffusion of matter we observe can be sufficiently conserved in their entire systems under the excessive amount of radiation they must experience in the clear cold regions of space. So that we must conceive that if the cometary state depended upon a force subject to such radiation, gravitation being constantly active within the system would, within a moderate period, reduce the comet to meteoric matter, which, owing to its reduced dimensions as solid matter, would after a period remain in this state, and become invisible to us, unless projected very near the earth.

179. Further, if we assume the comet to be entirely gaseous matter, we can scarcely imagine a degree of internal heat in the system sufficient to render this, that is the hydrocarbon portion of it, visible, neither can gaseous matter reflect the solar rays. Therefore, upon the whole, we are led to consider the gaseous condition as highly improbable to account for the observed tenuity, and at the same time the illumination of the whole comet as a visible body. On the other hand, it does not appear improbable, with respect to certain comets, that sufficient heat is maintained in the nucleus (in some cases, perhaps, from passing very near the sun) to render this visible in itself, and sufficiently so also to illuminate the surrounding matter of the comet and tail to some extent within a certain distance from the sun.

It is not necessary even to consider the nucleus a solid or liquid body. It may be a rotation-system composed of many parts reflecting light and yet transparent through the extent of the interspaces.

180. The conditions under which electricity of one sign,

\+ or −, could act within a unit system of matter such as a comet, are difficult to define; electrical phenomena depend generally upon the tendency to establish equilibrium, but with our limited knowledge we cannot say a single form of electrical energy in a system is quite impossible. If possible, it may place diffusion of finely divided matter in equilibrium with the reaction of gravitation for decentralization of matter in a comet. At the same time, with electricity we stipulate a form of energy which is exhaustive, particularly if light is produced, in the same way that heat is exhaustive. So that we cannot imagine its conservation in the same manner that momentum is conserved in matter when moving in a frictionless medium. Therefore, all that we can say is that electricity probably takes a part in the phenomena of comets, but this would not account for their symmetrical forms.

As regards illumination of the comet, the probability is that the electrical action present is a phenomenon exterior to the general cometary mass, exactly of the kind herein proposed for the illumination of the condensation of pneuma to nebula, by electrical discharge (§ 42).

181. We may now consider the third condition proposed:— *That the tangential action of the revolution of the outward parts of a gravitative system about a centre or focus may separate these parts proportionally to their velocities to any degree of tenuity.* In this proposition we may observe that, if the cometary volume is maintained by the revolution of its outward parts moving in elliptic orbits, we have conditions that bear a strong analogy to the motions of planetary bodies. Thus, so far as we know, all planets and systems that we can observe are in revolution, both in their own masses, and in the attendant parts of their systems (satellites, rings). Therefore by analogy, taking the comet to be a part of the solar system, we must assume that the *outward parts of the cometary system* are in revolution. And as the satellite may revolve at any distance from its primary in an elliptic orbit, so

also may any of the outward parts of a comet, however small, revolve. Further, if this motive system is assumed for the comet, we have the initial energy in the system conserved, as there is no loss as with heat radiation, unless we imagine resistance by a surrounding medium, of which we have no evidence from other planetary or cometary phenomena.

182. If we admit the conditions first suggested as regards heat or electricity to be in a certain degree active ; by this the nucleus of a comet may probably be heated or electrified liquid or solid matter in unit mass or compounded of many meteorites surrounded by a gaseous envelope, although this heat can scarcely be imagined to be sufficient to maintain the large outward cometary mass of the tenuity observed. Nevertheless, the nucleus would partly illuminate such exterior parts as I have suggested are in revolution about it, but in this we have clearly the necessity that such parts to be visible should be *solid* or *liquid* matter and not gaseous. Under this condition also the sun would illuminate the entire system.

In comets that pass very near to the sun it is presumable that through the great heat they receive near perihelion, any ordinary matter with which we are acquainted, and which might form part of the complete comet, would be reduced at the time to vapour or gas, in some cases possibly by explosion. This being the case, about perihelion, by radiation of heat received into space from the gaseous outward parts condensations may occur in these parts about separate centres, again developing heat and electricity. The units of condensation may form sphericles under the same conditions as clouds are formed in the earth system. These would appear in mass as cloud, and in this state would traverse the orbit of the cometary system, forming parts of the coma and tail. If the head of the comet was in rotation, the condensed units would be in revolution at any extended position therefrom.

Now, as regards the magnitude of each solid separate part

or particle now suggested as an outward revolving part of the cometary system, this may be as small as we like to conceive it, assuming its mass sufficient to reflect a *ray of white light*. For the existence of such minute cometary matter, we may possibly find some analogy in the system of small meteors which revolve in elliptic orbits about the solar focus, and which are sometimes brought within the attractive distance of the earth, evidence of which is further given by the cosmic dust discovered in snow by Prof. Nordenskiöld *.

183. A form of particle which may be probable as a result of uniform condensations is that of a smooth bright metallic or vitreous sphere, which is covered with a permanent gaseous condensation, of hydrogen or a light hydrocarbon, which, under the sun's influence, may produce a gaseous envelope in the same manner as the earth is surrounded by its atmosphere. The refraction of the gas intensifies the sun's heat and light upon it. Refraction and reflection will carry over part of the light to other such globes at greater distance from the sun. Further, when the comet is sufficiently near the sun for its heat to render the gas, if hydrocarbon, incandescent in the purely isolated state, the comet may possibly become self-luminous.

The change of state from the solid or liquid to the gaseous will also develop electrical conditions which may render its matter temporarily luminous when approaching the sun or receding from it by the effect of after-condensation.

184. If the exterior cometary matter in separate particles, as proposed, is in rapid revolution in very elongated elliptical orbit round the nucleus of the comet, which I assume is necessary to maintain the extent of what we may term the cometary volume of the tenuity observed, then this revolving matter may be projected in elliptic orbits, either as separate particles (satellites), or more probably in accumulations or

---

* 'Voyage of the 'Vega'.'

series in one or more connected gravitative systems in the form of more or less perfect rings or bands, the orbits of which, about the head of the comet, may cut the solar-cometary plane at all angles. I will now endeavour to trace diagrammatically what would be the action of such a system whilst moving in an elliptic orbit round the sun.

185. *Comet-tails.—Trains.*— Under the gravitative conditions proposed, the word *tail* will be entirely a misnomer, as the matter which forms the tail of the exterior parts in revolution about the nucleus is assumed to change position and become at another time part of the head. I will call the revolving parts of the head the *pericoma*, and the extreme of the tail the *apocoma*. For the entire comet except the nucleus, I will use the word *train*, which has been sometimes employed before.

186. Following the conditions proposed, that the comet is made up of separate parts moving closely together in rapid revolution around the nucleus, we must then assume, by the laws of gravitation, that the centre of gravity of the mass held by mutual attractions will be constrained to follow an elliptic path for its orbit. But the separate parts of the system which together form the *train* may describe elliptic curves about their *common centre of gravity, that is the nucleus, modified by their mutual attractions to each other, in combination with the attraction of the sun.* This may be taken in detail.

187. *Elongation of the entire Cometary Mass near Perihelion.*—Now, as regards the proper motion of the system of the comet constituted as suggested, we may consider that the velocities of its separate outward parts, as of all planetary systems including satellites, are such that the areas described by the radii vectores are proportional to the times—that is, in relation to their own focus, and so far as this system is undisturbed by other attractions. Further, we know nothing in astronomy of resistance by a surrounding medium; therefore, however small we assume the exterior separate parts

which are linked together by mutual attractions to form the train of the comet, these will separately maintain their velocities in relation to the nucleus consistently with the length of curves described by their radii vectores in equal times.

188. These premises being granted, we may next consider the conditions present of the relative attractions upon the separate parts of a cometary mass in combination with that of its superior focus the sun. For this we may first, to save the complication of superimposed motions, consider the comet upon the Swarm theory as an immense system of matter in separate parts, either as a number of separate meteorites or cosmic or nebulous matter, connected closely together by mutual attractions but forming an elastic system. The matter exterior to the nucleus is assumed for the present *not to be in revolution about the nucleus,* as our theory demands that it should be. Then suppose that the whole matter of the comet is moving in its orbit. Take this diffused mass at the position of aphelion, at such a distance from the sun that the sun's attraction would not materially disturb the arrangement of its parts. It will then be clear that as the large cometary mass (now assumed not in revolution) approaches the sun, each of the separate parts of the mass will be accelerated directly as the length of curve described by its radius vector in unit of time about the solar focus. Therefore, the forward parts traversing space towards the sun will move at this approach much more quickly than the following parts, and the entire comet, possibly globular at aphelion, will be enormously elongated into ellipsoidal form in passing very near the sun. By the same laws also acting inversely in passing from the sun, its mass will be contracted. This principle is shown by fig. 16.

Let S be the sun, and *a b c* three particles of matter in the train of the comet which is revolving in an orbit around the sun shown by the surrounding line. Let *a* be a forward particle, *b* a central particle, and *c* a following particle. Then will the velocity of *a* be greater than that of *b* at the period of approach towards the nucleus, and *b* greater than *c* proportionally to the lengths of curve described by their radii

Fig. 16.

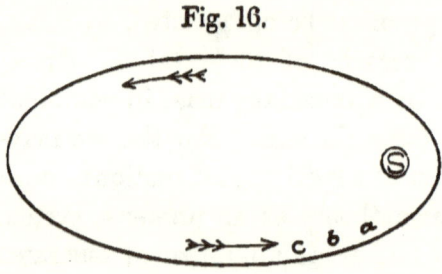

vectores in unit of time; so that as the comet approaches the sun its mass system or volume will be elongated in space in proportion to the excess of attraction to its forward parts by acceleration of gravity over its following parts. And under these conditions the greater the eccentricity of the orbit, that is the greater the acceleration at perihelion, the greater the length of the train of the comet. So that the outward form of the comet would in degree, so far as present conditions are considered, resemble its own orbit in form.

189. Under the above conditions, we may observe that, upon the whole, in considering the cometary mass as not being in revolution, all the parts in any plane would follow each other in elliptic orbits about the sun; and although the tail or following part would constantly increase in length in approaching the sun, it would not *diverge* from the sun in passing near it, or the separate parts move out of their separate solar orbits as gravitation units. Therefore there must be present, as generally admitted, other conditions to account for the divergence which is universally observed. In this proposition it is assumed that it is possible for the outer parts of the comet at greater distance from the sun to possess the same orbital velocity as the inner parts moving nearer

to it, so that the figure of the comet may be conserved, as erroneously assumed by others in the discussion of the Swarm theory, in opposition to Kepler's third law.

190. *Orbits of the outward parts of a Comet. Focal Point.*—We may now follow the conditions just proposed, and assume that the comet resembles a planet surrounded by a connected revolving system of matter equivalent to a system of satellites, or to Saturn's rings, this revolving matter being meteorites, dust, or nebulous matter, and that the revolving mass (which I have denominated the train) will not only be sensitive to the attractions of its own parts and its nucleus or the centre of inertia of the system, but also to the attraction of the sun at the same time. Then, as the sun's attraction will not be *linear with the major axis of the comet's mass* in its assumed elongated form, the outward parts of the train, which are projected forward of the nucleus during their rotation, will be accelerated and be drawn towards the sun, and if the motive direction of these is inwards towards the sun, as it will be in the exterior part in revolution in the solar-cometary plane about the nucleus, these will be *drawn towards the nucleus also*; and in the smaller arc thereby induced by increased attraction will have their velocity accelerated (Kepler's second law). In this manner, although the matter of the train will be elongated in space by the differences of velocity of its parts in passing over separate portions of the curve described by its radii vectores ; still the nucleus of the system will maintain a position at the forward focus of the orbit of the train, and parts of the train will be induced to move in an orbit round its nucleus, or the centre of gravity of the system, which will closely resemble that of the superior orbit of the entire comet in moving round the sun.

This matter may be better conceived by reference to the diagram fig. 17. Let S be the sun, C the nucleus of the comet, *a* a particle moving in its orbit in the solar-cometary plane, shown by an elliptical outline. Draw a line through the centre of the sun, and through the nucleus of the comet to represent the linear direction of gravity on the orbit

Fig. 17.

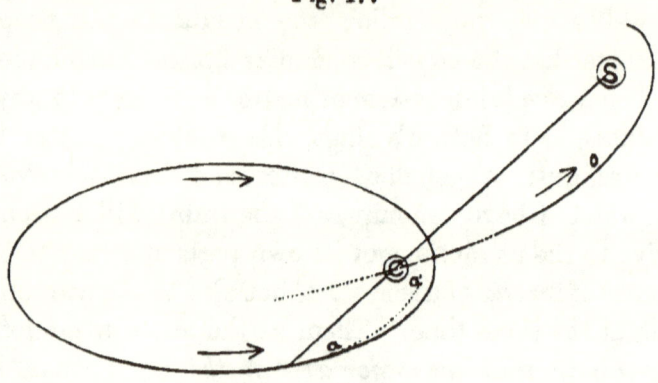

shown by the outline. Then will the combined attractions of the sun and the nucleus exert greater attraction upon a particle in this position *a* as the comet approaches nearer the sun, than was exerted upon it at the same position of the orbit by the nucleus only, and such attraction will cause the particle to move faster and nearer to the nucleus, or inwards in a direction from *a* towards the point *a'*. In such a position its forward radii vectores in relation to the comet's centre of inertia will be closed, and its velocity increased proportionately to the lengths of curve now described by its smaller radii vectores. It will thus also *maintain the nucleus at its focal point* within the comet, although the cometary mass be elongated by the increased velocity of its forward parts at pericoma as shown above.

191. *Direction of the Cometary Train in Relation to the Sun.*—By the above conditions it will be seen that the separate particles of the train of the comet will be actuated by forces which will be the resultants of the combined attractions of the sun and those of their own proper nucleus, the pericoma of the train, if this principle is admitted, moving

about its head will come constantly in the direct line of attraction between the sun and the head, and will constantly change the direction of the attractions upon the parts of the train. The particles of the train will also be subject to the attractions of their own masses upon one another; and if the train is a combination of dust or meteorite ring-systems as suggested, these attractions will still further modify the conditions given. The principle now offered may be best described diagrammatically by fig. 18, for which I will again take the position of revolving matter on the solar-cometary plane only.

192. Thus:—Let S be the sun, C, C', C'', fig. 18, three positions of the nucleus of a comet moving in the orbits Y Z, Y' Z', and Y'' Z''. Let $d\ e$ be two particles of matter which form part of a continuous attractive series constituting the train. When the nucleus is at C, and moving in the

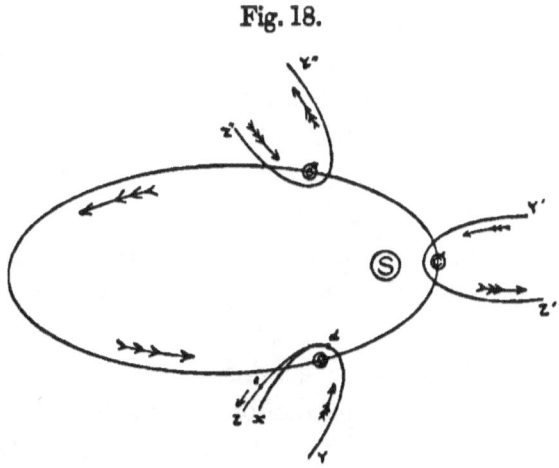

Fig. 18.

orbit Y Z, then will the continuous series of particles arriving at d be not only actuated by the attraction of their nucleus, but will be accelerated by the attraction of the sun directly linear to their motive path in the direction d, as before shown a to a' in fig. 17, and all following parts as they arrive will be mutually attracted in series in the same direction. Therefore the particle d, fig. 18, will be both accelerated, and be drawn at the same time towards its nucleus C, and by the curve of its smaller radius vector it will pass nearer the nucleus with greatly accelerated speed.

193. Now, as regards the attraction of its nucleus and the velocity engendered in particles moving about it by the attraction, the velocity would be maintained, varying only in inverse proportion to the squares of length of its radii vectores onwards to e; but from d to e the sun's attraction would *still draw the particles towards the cometary nucleus*, so that throughout perihelion passage the curvature would be in a certain degree maintained as at perihelion, directing the particles of the train thereby towards the point x. The apparent effects of this will be that the whole mass of the train considered as an elliptic mass in revolution will be, as it were, pulled forward in the direction of its rotation, and whirled round in space *so that the position of the sun's centre may be maintained constantly linear with the least radius vector of the parts of the train that pass nearest to it*, and the whole train of the comet, although describing an ellipso-epicycloidal curve, will be as far as possible constantly symmetrical about the nucleus; the orbit of the train thereby changing to make this possible from the positions Z Y to Z' Y', Y'' Z'' when the nucleus of the comet arrives at positions C, C', and C''.

Under the above-stated conditions the revolving matter of a comet passing its perihelion will possess much greater absolute velocity than the centre of gravity of the comet. Therefore a comet may pass very near the sun's surface with momentum too great from this self-rotatory velocity for its matter to be materially disturbed by slight resistance of attenuated matter, if such exists about the sun.

194. *Widening and Curvature of the Train by crossing Orbits.*—One other point to be considered under the conditions

given above is that the newly placed pericoma at every change of position in relation to the cometary nucleus, as it becomes directed towards the sun, will influence the parts of the train only in proportion as they are accelerated. Therefore, the parts of the nucleus which have just passed the pericoma will have the accelerative force due to the sun fully impressed upon them, whereas the parts arriving there will not be so fully impressed, but will retain a part of the force due to their initial velocity in relation to the previous position to the nucleus at the last time they passed between it and the sun. Therefore, the following parts will not maintain quite a symmetrical position with respect to their future pericoma, but will lag behind this position, so that the general path of projection of the train will be constantly of greater curvature upon the exterior of the orbit than in the parts nearer to the sun.

195. If the matter in revolution about the cometary nucleus revolve in different periods according to Kepler's third law, as the planetary matter about the sun does, then the orbits

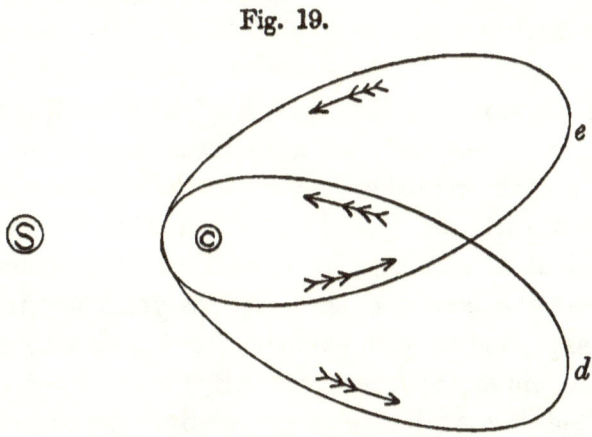

Fig. 19.

of the parts of least period of revolution, which will be nearer to the nucleus, will appear to lag less than the outward parts of longer revolution period about the nucleus, so that by this means a crossing of orbit will occur, producing either a *widening or opening of the tail.* Under these conditions also possible collisions may occur, producing complicated effects impossible to follow here. Fig. 19 represents the normal conditions in the solar-cometary plane where the crossing orbits are assumed to be conserved.

196. Let S be the sun, C the cometary nucleus, d represent the orbit of a particle of early revolution, e the orbit of a particle of later period. After perihelion of the entire comet in relation to the sun, this system of forces, by the changed dispositions of attractions, will act inversely.

197. *Formation of a new Head or Pericephalion about Perihelion Passage of the Comet.*—Under the above conditions, independently of the complication of motion of cometary matter moving in orbits not in the solar-cometary plane, the conditions of which have not been separately considered, incidental phenomena must occur which will, more or less, disturb the general conditions thus:—As the sun causes acceleration to the parts of the train in approximating their pericoma in relation to the head, and retardation to these parts in leaving it, the matter of the train will be more condensed round the head in the point of pericoma of the system, and thus form incipiently a new *head*, which will react, by its gravity, as a secondary or false focus, and in its turn, by maintaining the centre of gravity in the cometary orbit, will tend to disturb the position of original matter that formed the original focus. At the same time, if the comet passes very near the sun, the head, by the great heat it receives, will be expanded or even exploded; so that it may become a less dense mass; or even new heads may be again formed forward of this with a general relative disturbance of the internal gravitation in the parts of the cometary system, the conditions of which are too difficult to follow.

198. It is at this point where, as pericomic matter increases in density at the head of the comet, and heat forces, and

possible development of electricity, cause great increase of internal elasticity, the orbits of train-matter may be separated by explosion, so that a comet may be divided into two or have a large part of its train detached from its own focus, one part to be retarded in its orbit as much as the remaining part of the comet will be accelerated mass for mass. Or the complexity of orbital conditions induced may separate a part of the train about perihelion where solar action is most intense.

199. *Some General Conditions.*—As to certain appearances of comets under the conditions proposed, which will influence the spectra obtained therefrom if the nucleus of a comet is a heated or electrically excited mass, which it may become by the collisions of parts of the train in revolution near pericoma, or by heat derived from the sun. Under the same conditions the train in passing pericoma may be also heated or electrically excited sufficiently to become luminous or phosphorescent. This luminosity will gradually die off in the distance, so that possibly the parts of the train of a comet in aphelion are seldom visible. From which we may conclude under these conditions that *visible comets* are generally only the parts of the cometary system that are nearest the nucleus. In the comet of 1882 (Wells) I was anxious to observe whether any appearance could be detected of orbital projection beyond the extent of visible tail, which could be possibly observed by some obscuration of part of the faint light from celestial space circumscribing a curved elliptical space. This partial obscuration I fancied I observed on the 23rd October at 5 A.M. It was also observed as a darker part behind the visible tail by Mr. C. J. B. Williams at Cannes, France*, as also by Mr. Henry Cecil at Bournemouth †.

200. Upon principles here discussed the cometic matter in revolution about the nucleus causes each particle to be

---

* Nature, Oct. 26, 1882, p. 662.   † Nature, Nov. 16, 1882, p. 52.

illuminated by the heat or electrical excitation engendered through constant collisions, also by the heat of the nucleus, and by that of the sun on the side turned towards it. Therefore the light we receive by reflection will be proportional to the amount of illuminated surface of the revolving particles visible to us, and the direct light that of incandescence. In this manner the nucleus itself may become invisible or dimmed by eclipse of revolving incandescent matter about it. On the other hand, the light received directly from the nucleus will be proportional to the open spaces only between the outer revolving parts crossing the field of light. From these causes it is possible that some of the most remarkable appearances of projection of cometary matter suddenly from the nucleus through immense distances may be merely the lighting of the distant parts of the train of cometary matter through interspaces of the central system upon dispersed matter which, although present, was previously invisible to us.

201. The matter of some comets, of which Encke's is perhaps one, may be a carbon-hydrogen compound, which in the distance of space, as at aphelion, may be condensed to particles of solid matter, but in nearing perihelion may be again converted into gas by the heat of the sun with development of incandescence through electric excitation sufficient to render it visible. This matter may again condense in passing towards aphelion, the orbit position of the separate units of the system remaining the same with the electrical effects of change of state—the apparent outward visible volume of the comet varying according to the conditions observed.

CHAPTER X.

The Earth.—Considered in evidence of former Nebular Conditions. — Its internal Fluidity.—Tidal Friction.—Change of Figure due to Rotation.

202. *The Earth* may be considered as a model planet whereon we are able to observe the evidences of exterior conditions which are due to phenomena that have acted upon it to produce its present form and constitution. In this study we may possibly approach the conditions which also ruled in the formation of all the dense planets interior to Jupiter, of which we can possibly obtain no further evidence of structure than that apparent upon the surface of Mars and the moon.

203. To assure our premises for earth-structure it will be convenient briefly to recapitulate some general propositions that have been already discussed, which we may possibly accept as data for certain factors of early formation. The most important of these are :—1. That the solar-planetary system during condensation possessed more or less nebular matter projected about its equatorial zone, as represented diagrammatically by the discoid form in fig. 9, p. 67. 2. That the tenuity of the planet-forming system interior to Mars upon the condensation of the sun's volume, was too great to support the nebulous concrete state at a period when the exterior solar nebula was falling below its critical temperature (§ 112). So that within the orbit of Mars local separate condensations were formed at first of the more refractory matters

widely exterior to the earth's orbit. 3. That at the early period when the intra-Mars local condensations were forming the earth's nebula was still attached to the sun (§ 146). 4. That the exterior local condensations of matter at an early period were drifting sunward into the earth's nebula and towards its orbit position in moving under the resistance of the surrounding less refractory nebulous matter. So that the earth during its formation at no time extended in an entirely uncondensed nebular form so far as the orbit of Mars. 5. That the earth's formation being partly but not entirely due to the condensation of gaseous matter as assumed herein of the planets Jupiter and Saturn a denser system was produced.

204. Under the above-stated conditions we have the suggestion of two large factors of earth-formation:—An interior nebular system about the earth's orbit of purely gaseous elements, which were sinking at the period of formation to a critical point of temperature and impressing their virtual momentum upon the earth in the direction of its rotation, upon principles already discussed § 128, and an exterior condensation system of meteorites or planetoids which were projected into the earth's nebula. These, as bodies moving in nearly free orbits, that is, under slight resistance of the residual nebulous matter, drifted into the earth's nebula in spiral orbits that upon contact with the then forming earth may have produced effects which continue to be evident upon its surface.

It is necessary in this matter to insist upon the continuance of nebular conditions during the greater part of the period of the formation of the mass of the earth, as such conditions alone could have produced its present direction of rotation, as it was shown originally by Descartes and Laplace (§ 7), and with equal clearness by M. Faye (§ 141), that if it were formed from an entirely discrete system of matter moving in free orbits its direction of rotation would be the reverse of what it is.

If we omit from consideration the projection of discrete matter into the earth's nebula, which would carry with it certain elements of relatively retrograde momentum according to the theorem of M. Faye, the necessary extent of nebula, if this alone was active, may be inferred by taking the formula $2\pi r_2 D = 2\pi(r_1 - r)$, from which we find $Dr_2 + r = r_1$, D being the days in a year, r_2 the radius of the earth, r the radius of the earth's orbit, r_1 the outer radius of the nebular ring, which by calculation to produce the present rotation is found to be of about 1,000,000 miles greater radius than the earth's orbit.

205. If while the earth was moving at nearly orbital velocity there were projected upon it certain factors of discrete condensation, giving a momentum the reverse of that due to direct condensation of gaseous matter, the radius of the original gaseous matter would have to be increased in proportion to the momentum of the discrete matter in order to account for the present rotation period.

If we accept the conditions proposed above, they entail certain relative consequences of which we should expect to find evidences in earth-structure.

206. *Firstly*: If the earth were condensed from a gaseous system, this condensation could not have been effected without producing an intense heat in the condensed matter, as it may be supposed to have resembled our sun in constitution at an early period. Further, such heat as would be produced by gaseous condensation must have also rendered any discrete matter which may have been projected therein liquid.

In the formation of oxides upon the earth, if these were produced from elementary matter, the only probable condition, they must also have produced great heat during their oxidation. At the same time, as such oxides are lighter than their metallic bases and are non-conductors of heat, they must have rested upon and covered the surface and conserved the central heat. Therefore we demand, in the first place, upon these conditions, evidences of a highly-heated liquid interior.

207. *Secondly* : If the planetoids, large or small, produced by the condensation of matter exterior to the earth's nebula were incorporated therewith when the earth became a liquid planet, then such matter within or upon the earth's surface might possibly remain geologically evident. Particularly such planetoids as may have drifted to the earth's surface when all the denser matters of its nebula were condensed. Under these conditions it remains probable that some land-areas of the earth may show evidences of this discrete form of condensation. These conditions are so far actual that we have still planetoids, that is, meteoric matter, falling to the earth.

208. The nebular conditions which upon the theory herein proposed would have been constantly active at an early period of earth-formation require consideration in detail. They will therefore be deferred to another chapter with the exception of the argument upon which they must be supported of the entire internal fluidity of the earth. This also is necessary for the consideration of the effects of later discrete projections upon the earth to be discussed in the following chapter, as I suggest that these discrete projections were coincident with nebular condensation and may remain, at present, evident in land-formation.

209. *The Internal Fluidity of the Earth.*—This could not be questioned for a moment if the evidences of observation were alone considered, but in the contemporary science of any period we have generally in popular learning a tendency to depart from concrete observations directed to consider some hypothesis or isolated fact, which is elevated to predominance above all the natural conclusions of otherwise universal observations. I remember, when a cool sun was the prevailing theory, arguing with a professor that the intense heat of that body was manifest in many ways. His reply was that that was nothing to the purpose if the solar spots were found

to be hollow places or cavities in the sun's surface; as they were dark they must be cool; and science must take account of every phenomenon. If experience teaches us anything, it is that theories change with every phase of science; so that our safety, if we are seeking truth, lies in taking the mean evidence of all relative phenomena, not of any single phenomenon which we may or may not perfectly comprehend.

The most important evidence of the former nebular condition of the earth besides its direction of rotation is to be found in its interior condition; for it is certain, as just stated, that the matter which forms the earth, if it previously existed in a gaseous or vapourous state, must have been condensed from this state to a liquid with great development of heat. The density of the system also clearly indicates that the internal matter, if at a high temperature, is most probably metallic. Now as it is the property of metals to alloy and also to conduct heat, it is in the highest degree improbable, as sometimes supposed, that any part of a uniform system of condensation, as that of the earth, could be for ages intensely heated in some interior parts, and cool or solid in other parts. It therefore becomes a rational condition of the nebular hypothesis that the interior of the earth was, or is, in a highly heated uniformly liquid condition.

210. If the earth at an early period condensed refractory metallic matter to form an intensely heated globe, while this was surrounded by less refractory matter, and particularly by oxygen, near the surface, then the globe would be formed mainly of the refractory matter, and be covered with the oxidized matter upon its further condensation. The oxidized matter must necessarily have rested upon and outwardly covered the metallic matter, so that possibly at a very early stage of condensation there was a metallic uniformly heated globe covered with a coating of non-conducting oxidized matter as we find it at present, although this coating at an early period would be very much thinner than it is now.

As the earth cooled the coating must have increased by condensation and protected the earth more and more from radiation of internal heat until it was possible for water to rest upon the earth; so that the heat of the interior must have been protected by non-conducting material, highly heated from chemical combination, in such a manner that it could only very slowly radiate its heat into space.

211. We now arrive at an important point. Is the interior of the earth fluid? From mathematical physics alone in the accepted theory of the tides, Lord Kelvin, our greatest authority, says that it is solid, or the tides would not present the great variation of height of surface which we witness. This matter is repeated with authority in the recent compilations of Lord Kelvin's works[*]. There is one point of this theory at least which astronomers ought to be able to solve. Lord Kelvin shows that the necessary result of tidal friction is that the earth's rotation must be decreased by a total value of twenty-two seconds in a century; and the establishment of this as a fact might tend to prove the certainty of the effects of tidal friction upon the earth, and at the same time possibly, if we had no other direct observations to consider, of the earth's solidity. If we depart from physical theory and follow the evidences of astronomy by observation, wherein the data are results of experience not liable to receive much correction from change of theory, we have undoubted authority from observation that the rotation period of the earth, considering the moon and Mercury particularly as time-keepers, has not decreased by a single second in a thousand years.

212. If we follow the evidences of geology founded upon observation, the solidity of the interior of the earth appears to be quite impossible unless we assume an unknown form of matter which does not become fluid at a high temperature. We need only cite a few instances to show the improbability

---

[*] Prof. O. J. Lodge. Review in 'Nature,' July 26, 1894.

of the rigid solid earth proposed by Lord Kelvin. In every considerable volcanic eruption, even within modern times, the matter that issues from the earth is at a white heat; and although this may be reduced to dust, as in the Krakatoa eruption, the microscopic examination of the dust shows clearly by structure that it was formerly, or just before its issue, at this heat and in a liquid state [*]. We find also that this matter may again be reduced to liquid at a heat much less than that at which it was thrown from the volcano. The mass of matter projected from Krakatoa is estimated to be equal to about 22 solid miles. The eruptions of other mountains in Java were of much greater mass. The liquid lava that issued from Skaptar Jökull in Iceland, in 1783, is estimated at 21 cubic miles. The lava of the Snake river, Idaho, North America, that has issued in tertiary times is not of much less volume than 500,000 cubic miles. We have also upon the surface many thousand cubic miles of lava in Abyssinia and in other parts that has certainly issued at a white heat. Further, the axes of our mountain chains are built up of plutonic matter the structure of which shows the evidence of former intense heat with extrusion of liquid mineral matter. This is evident although the surface-matter is often dislocated and rearranged so as even to include superficial sedimentary rocks, the deep-seated felsitic rocks seldom intruding beyond the base of a volcanic chimney. The volcanic and sub-volcanic matter, therefore, upon the earth's surface, which has evidently been in a liquid state, amounts to many millions of cubic miles. When we consider the wide distribution of this matter, its nearly uniform mineral structure, and its present almost constant issue, as in Kilauea and other volcanoes, we cannot but conclude that it formed, and now forms, a part of a general system whose intense heat

---

[*] See author's paper, "Krakatoa Dust," R. Met. Soc. Quart. Journ. vol. x. p. 187 (1884).

is derived from the uniformly heated state of the central matter of the globe.

213. The evidence of the thermometer, which shows that there is a general increase of temperature with depth of about 1° Fahr. for every 60 feet, where freedom from the influence of percolated water permits this measurement to be made, gives an increase of temperature of 88° per mile; so that at about 200 miles we should have, at this rate, a temperature of 17,600°, which no known body could bear while retaining a rigid state. It is not, however, necessary to consider a constant increase of heat if the central volume is metallic, as herein proposed, for a good heat conductor would distribute heat equally in the interior. The following suggestions may be offered as regards the internal liquidity of the earth.

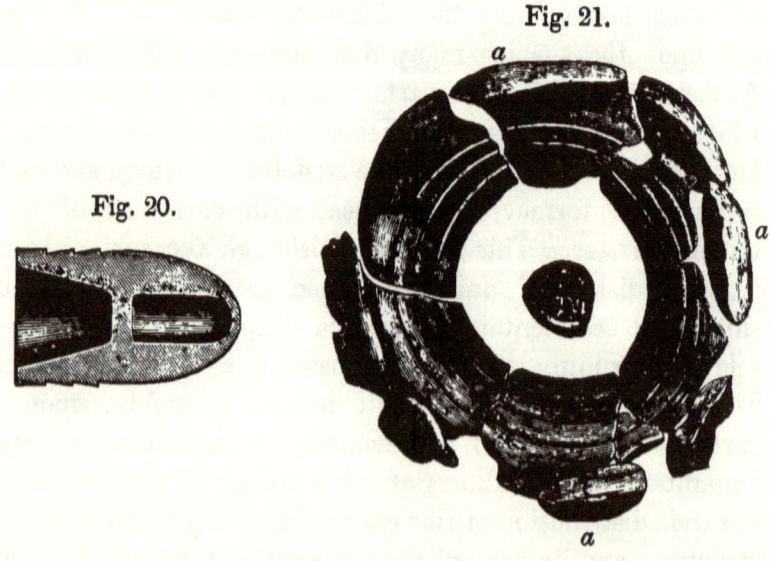

Fig. 20.

Fig. 21.

214. Is the assumption of the solidity of the earth necessary to account for tidal action? Having devoted some years to the consideration of motion in fluids, I find that the mobility of liquids depends greatly upon the freedom of surface, which adapts it to offer certain forms of accommodation for motion

THE INTERNAL FLUIDITY OF THE EARTH. 155

for which time is required. So that a dense liquid held in equilibrium by surrounding pressures, such as we may imagine the interior of the earth to be, offers considerable resistance to deformation through molecular friction in moving from a state of relative rest until certain forms of accommodation can be brought about, which are nearly impossible in a close system. In firing an Enfield rifle bullet directly upon the surface of water, the point of the bullet does not pierce the water, but the water pierces the bullet, and reduces it to a thin conical shell *. Fig. 20 shows a full-size section of the bullet. Fig. 21 was taken directly from this bullet fired from an Enfield rifle normal to the surface of water through a thin piece of bladder in the end of a deep tank. The point of the bullet was painted red, and left to dry. The colour upon the point was found spread out to the extreme circumference, where it formed a curled-up edge, as shown in the figure, $a\,a\,a$. The centre of the figure shows the pressed out diaphragm of the bullet, which was reflected from the surface of the water, so that it forms no part of the scheme.

The velocity of the earth's surface at its equator is much greater than that of the bullet; therefore we have to consider accommodation in the waters of the ocean and of the liquid interior for displacement in relation to time, in order that it may possess a certain form of possible motion of accommodation. This presents a difficult problem, the factors of which are, for the greater part, quite unknown. If, on the other hand, the deformation under the moon's attraction is considered to depend upon the elasticity of the system, then a solid is quite as elastic as a dense liquid under great compression, so that neither solidity nor liquidity could be inferred from this cause. Gold or steel in a solid state is much more yielding under pressure in confined space than water, as is proved in

* 'Fluids,' p. 187.

the coining of metals, in contradistinction to the compression of water in the hydraulic press.

215. Judging from the constitution of meteoric matter that has fallen upon the earth, assuming cosmic matter in a certain degree general, the centre of the globe would be largely composed of nickeliferous iron. As regards this matter, we have not been able to produce any temperature in our furnaces high enough to melt it, so that assuming this formed the larger part of the interior of the earth upon which the surface rocks rest, the white heat of volcanic matter as it issues from the surface of the earth, taken as an index of the interior temperature, would only be sufficient to raise pure iron or nickeliferous iron to a stiff plastic state, as iron in forging offers very great resistance to change of form under great surrounding pressure.

216. It is certain that the earth's rotation upon its axis is combined with that of the rotation about the centre of inertia of the earth and moon; therefore there must be a swing in the free surface of the ocean, due to its plus and minus daily and monthly rotation-velocity in relation to its position with respect to the moon, which must produce tidal action in the free surface water. But how far this plus momentum in one part is compensated by the minus momentum in another may be a difficult problem to solve. One must, however, feel in this, as in other instances with which history furnishes us, that in the intent observations and calculations of certain actions we are liable to lose sight of the reactions that are not superficially evident or may be unknown. So that it becomes a question whether the motions within and about the earth moving in its orbit in frictionless space seriously affect its general momentum of rotation taken in a wide astronomical sense, so that we may infer its internal condition therefrom. Certainly, taking the matter in its entirety, we cannot suppose it to deviate greatly from Newton's Third Law, Cor. iv.: "The common centre of gravity of two or more bodies does not

alter its state of motion or rest by the actions of the bodies among themselves."

217. *Effects of Tidal Friction.*—Taking account of the suggestions offered in § 153 for comparing the revolution of the moon with the rotation of the earth upon nebular conditions, the effect of tidal friction upon the present rotation of the earth must have been very small, the proof of which is quite evident in early geological stratification. That there have been certain effects is suggested to be made evident by the present acceleration of the moon of 10" or 11" in a century. This effect was suggested by Kant and afterwards worked out mathematically with varying results by many eminent astronomers. Laplace accounted for the acceleration by decrease of ellipticity of orbit. These calculations being revised by our late eminent astronomer and mathematician Adams with greater precision, reduced the Laplace factor by 6", leaving a residual 4" or 5" to be accounted for by tidal friction or *some other cause* *. Whether this residue may be a recurring differential from astronomical causes remains to be proved. If we take it to be a constant of acceleration due to tidal friction, then subtracting the Laplace-Adams factor, and assuming the earth's angular rotation-velocity to have been retarded in the same ratio as the moon's angular revolution-velocity was accelerated, and considering 5" per century as the mean acceleration, we have a period of about 12 millions of years for the time that the angular velocities of the earth and moon were equal; that is, the condition assumed herein of the purely nebular state before the separation of the earth and moon, which must have been followed by a long period of time before there could have been any geological conditions that could remain evident in the structure of surface rocks. This short period can, therefore, scarcely be discussed if we take any recognition whatever of geological evidences.

* Phil. Trans. vol. cxliii. p. 397.

218. *Change of Figure of the Earth due to Rotation-velocity.*—If there are sufficient scientific data to assume that our present day is longer than formerly, of which there appears to be a probability only within very narrow limits, then this, so far as concerns the present investigation, points to the conclusion that the mean symmetrical figure of the early earth as a spheroid of revolution must have been changed by this condition and have been at the time of greater rotation-velocity more oblate. Upon the nebular condition just discussed, the difference could not have been great. The effects of contraction of the equator under decrease of velocity, if we may take it at the extreme value accepted by some scientists, was ably discussed in a paper read by Mr. Wm. B. Taylor * before the Philosophical Society of Washington, May 3, 1885.

219. Mr. Taylor takes the very extreme condition proposed by Prof. G. H. Darwin of a rotation period of six hours, which must have produced an elevation of land at the equator one-tenth greater than at present, assuming the earth to have been a true spheroid of revolution. He finds that this would make the equatorial radius 4359 miles and the polar radius 3291 miles only. The pole would be therefore 658 miles nearer the centre and the equatorial protuberance 396 miles higher than at present. Mr. Taylor suggests that this would account for all the crumpling and inclination of strata and the elevation of mountain-chains. It is very doubtful whether it would do so; the mean inclination of strata at all depths is probably not less than 20°, and this could not be produced even with one tenth the equatorial contraction, supposing the inclination in the past constantly increased instead of oscillating locally up and down during all geological time, which must have been the case from local volcanic disturbance that is quite evident in extreme cases by some strata being quite over-

---

* American Journal of Science, 3rd series, vol. xxx. p. 249.

thrown and inverted. Nevertheless it may be much more a *vera causa* than the contraction-theory of the late Robert Mallet. It may have been a cause of elevation of land in the tropics as an early condition, but it must have been materially modified by the condensation of nebular matter at the poles which I shall propose further on.

220. The effects of the rate of rotation of the earth upon its oblateness appear to have been first suggested by the Rev. O. Fisher, and carefully considered; but he regards them as very slight. In his interesting work on the 'Physics of the Earth's Crust' he says :—" The friction of the tides, whether oceanic or bodily, must necessarily have diminished the rotational velocity and lessened the oblateness. The parts of the crust about the poles will have been subjected to stretching and those of the equator to compression. There is, however, no apparent reason immediately to connect the inequalities with this cause, for the continents do not occupy an equatorial belt, as they would do under this hypothesis, nor have the polar regions been free from the compression which all continental areas have experienced "[*]. This fact appears to me to be a sound objection to the whole hypothesis of a former rapid rotation of the earth, seeing that the known geological evidences of early stratification are directly opposed to it.

[*] 'Physics of the Earth's Crust,' Chap. xiv. p. 183.

CHAPTER XI.

SUGGESTED SUPERFICIAL CONDITIONS OF THE FORMATION OF THE EARTH, PARTICULARLY THOSE DUE TO DISCRETE CONDENSATIONS WHICH MAY HAVE BEEN FORMED PRINCIPALLY BETWEEN THE EARTH'S ORIGINAL NEBULA-ZONE AND THE ORBIT OF THE PLANET MARS.

221. *Formation of Land-areas by inclusion of Planetoids into the Earth's Surface.*—In this chapter it will be convenient, in order to simplify the subject, to take the conditions which were proposed *secondly* (§ 207) of the factor of earth-formation from planetoid matter, deferring the more important consideration of nebular formation until the next chapter. The fall of meteoric or planetoid matter which may have been formed originally by condensation in a part of the space-interval between the earth's original nebula-zone and Mars will be now considered.

222. Assuming the discoidal system of our sun's nebula shown in fig. 10, p. 85, we should have between the earth and Mars a thin attenuated nebular plane, in which local condensations would occur similar to that before suggested for the system of asteroids (§ 112). Under certain local conditions such condensations may have assumed any possible dimensions of mass according to the amount and disposition of the surrounding nebular matter and to its initial motion being directed so as to form a centralized system or otherwise. Upon these conditions any local condensation would form separately what we may term a small planetoid, or merely an isolated unit of impalpable dust.

The units of discrete solid matter formed within the space above defined, capable of composing any part of the future earth, must have possessed at the time of their formation a revolution-velocity less than that required for a circular orbital motion, according to the law of orbit. Otherwise they would have maintained their orbits and still circulate round the sun, a possible condition at present for many of these condensations. If the condensed matter was endowed with a velocity less than the above, as before stated, the sun's attraction would draw the early condensations into elliptical orbits; and consequently such as had a perihelion distance from the sun within the distance of the outer periphery of the earth's nebula-zone at the time must have fallen into this zone and have combined with it, through the resistance which such projection would encounter in passing into or through the nebula-zone. In this case it is easily seen that solid matter projected into the earth's nebula-zone must have combined motively with the momentum of the matter of the zone, and, if it retained its distinct solid condition, fall in spiral paths sunward, unless it was drawn by a stronger local attraction within the zone-ring or the nebulous globe of the incipient earth.

223. The period when exterior condensation would intrude into the earth's nebula-zone would depend upon many conditions. If its perihelion passage was entirely resisted by the zone it would enter a part of its system; but if the resistance was insufficient, on account of the small density of nebular matter, it might continue its projection with an orbit of changed ellipticity in any degree. If it arrived at perihelion within the zone-ring after the ring had condensed into globular planetary nebulæ or into a single nebular globe, its perihelion passage might strike or miss this nebula, or later miss the new-formed earth upon its further condensation in such a manner that there may be remains of intra-Mars condensations still falling as meteorites across the earth-orbit owing to want of former coincidence of

M

perihelion position with the earth's place at the time of internal conjunction.

224. Under the above-stated conditions, during the discrete condensation of intra-Mars nebular matter and its projection earthward, so far as this could overcome the resistance of the nebulous matter to continue its projection in a solid form, it would form earth-surface matter in combination with the nebular earth-zone which would still be forming. By this entire effect, the surface of the globe would receive a mixed condensation of matter upon its surface composed of the discrete matter, which would contain every form of solid elementary matter, and of the residual nebulous matter that was condensing at the period.

The above-defined mixed conditions of deposition correspond fairly well with actual observed conditions, as it is clear that we have no regular density system such as would be produced under the condensation of a purely nebular system. We have gold, copper, iron, and other dense metals upon the surface, which may at an early period have formed the central systems of discrete condensations from the universal pneuma before their projection to the earth.

225. In the small amount of exterior matter that has fallen upon the earth in recent times we have been able to recognize the presence of 23 elements, for the greater part those that generally prevail upon the earth, but in a few cases some of its rarer elements. They show upon the whole that they are condensations from an original universal nebula, in which the heavier as well as the lighter elements appear, iron largely predominating. The mean density of the entire mass of such meteors as have fallen to the earth in modern times is possibly not far different from the mean density of the earth, about $5·6$ times that of water, which should be the case under purely open nebular conditions of condensation from a universal pneuma system.

It cannot, however, be suggested that the presence of

heavy metals upon the earth's surface is entirely due to its reception of meteoric matter. There are other sufficient causes made quite evident in special cases. Metallic veins appear in many cases to be caused by the evaporation of metals from the heated interior, through fissures which were produced by the upheaval of primitive and later rocks. This evaporized matter is either in the form of pure or alloyed metals or of haloid compounds often modified by the presence of heated water and possibly originally combined in gaseous emanations. There is another cause to be proposed further on, namely the effects of the loading of ice at the poles upon the central system of the globe, causing projection of interior matter; so that the formation of land by the inclusion of meteorites may not be thought to be even a necessary condition of its formation, but only a very probable factor.

226. To recapitulate the condition proposed that possibly will agree with observation of geological structure :—We may assume that after the formation of Mars at the limits of the sun's nebular periphery, planetoids were again formed somewhat similar to our asteroid system existing exterior to Mars, from condensations due to the reduction of the temperature to the critical temperature of the sun's nebula within a thin peripheral plane, M, E, fig. 10, p. 85. We may assume that these planetoids were formed at first of the more refractory nebular materials. That they moved at first in their partially free orbits in spiral paths sunward under the resistance of the residual, more attenuated, or less refractory nebular matter that remained by its elasticity above the denser part of the nebula nearer the sun's surface. That upon further shrinking of the sun's nebula and fall of temperature in a more regular manner, the earth's nebular zone was first formed as a mass extending in volume much beyond the moon's orbit under conditions already discussed. That this condensation, which finally formed the earth, was set in rotation consistent with its mode of formation due to the

difference between the linear velocities of its outer and inner parts, which were moving at equal angular velocity, as before discussed (§ 146).

227. That under the above-stated conditions, the nebulous earth, being of much larger mass and condensing much later than the outer small planetoids, remained gaseous for a long time after they had condensed. That it became a liquid globe surrounded by a voluminous nebulous atmosphere at the time that the intra-Mars planetoids represented in miniature cold bodies similar to the present earth. That these planetoids after cooling possessed dense metallic centres, probably still highly heated, surrounded by oxidized metals and haloids, with water or, more probably, ice and air upon their surfaces. That the planetoids moving under the resistance of the still attenuated nebula surrounding the sun with their small excess of angular velocity, drifted slowly as regards inward motions sunward into the nebula of the new-forming globe until they were drawn to its surface and became finally incorporated about the perihelion positions of their orbits. The effects of the percussion again developing great heat, but not sufficient to convert the perfectly cooled planetoid projected upon the earth's surface, if this was of large mass again into a liquid state, or to bring the earth and the planetoid back to the form of a single nebulous globe.

228. Such systems of collisions as inferred above at every union of a planetoid, if this happened to be of large mass, might be considered to form at the time of contact a close binary system. The parts of the planetoid in contact with the earth would become highly heated by the effect of the collision, and sinking with its lighter coating into the liquid central globe of the earth would produce convection currents during its immergence, bringing the lighter liquefied surface matter into equilibrium of the mean gravitation system. In this process the lighter oxidized matter from the former buried surface of the injected planetoids being slowly floated up to

the surface of the earth would remain projecting beyond its mean spherical surface, so as to form a future land-area.

The extruded oxidized mineral matter of the planetoid would carry with it part of the metallic matter that forms the interior of the earth. Thus, through the friction of the operation and cohesion, such dense matters as gold, platinum, and other metals formerly in the core of the planetoid would be placed much above their mean gravitation position on the earth's surface.

229. The condition of our planet after the period of the last collision of a planetoid of any considerable mass would be most probably represented by an irregular spheroid surrounded by a nebulous atmosphere, still in a highly heated state from the temperature due to its original mode of formation and from that developed by the collisions of the planetoids from which certain factors of its land-areas were possibly derived. We may consider a particular case.

230. *Projection of a large intra-Mars Planetoid upon the Earth.*—We will assume for the formation of a continent for a special case that a cold planetoid came into collision with the earth at the position of South America, which was of a mass greater than this continent, measured above the level of the ocean. That the planetoid moved in a spiral path under the resistance of the earth-nebula in the same direction as the earth's rotation, so that it reached the earth with what we may term a very moderate percussion. That the earth at the time was a liquid metallic globe, covered only by a thin coating of oxidized matter. The effect of the concussion would be that the surface of the cold planetoid would be raised to a white heat; its water or ice would evaporate into the nebulous matter about the earth; its oxidized surface would sink at once into the liquid earth-globe and become liquefied itself. This lighter liquefied viscous matter of the exterior surface of the injected planetoid, composed mostly of silicates as in the earth-system, would be floated up slowly

about the periphery of the intruded mass, and the solid centre of the planetoid would gradually sink into the globe until it reached a position of gravitation-equilibrium and became a part of the interior metallic density-system of the globe. This could not occur, however, without much confusion and violent plutonic action within the central matter during the time that the lighter oxidized former surface-matter of the planetoid was reaching the permanent surface of the globe by convection currents. These currents would form special drifts and bring much of the metallic matter to the surface by tearing it away in collision, as it is a property of semi-fluid siliceous matter to bite into a metallic surface, as we find in the process of enamelling.

231. The final result of such a motive system as here presented would be the projection of a mass of vitreous rock above the earth's surface, entirely added to the earth's superficial system, forming a hollow plane, produced by the excess siliceous material floated up at the periphery of the included planetoid. The lowest part of the surface of the earth covered by the projected planetoid would be raised above the mean gravitation-plane of the earth's surface, the entire upheaved rocks being in equilibrium-excess of mass of the volume of the vitreous rocks upon the surface of the partially included planetoid. The effects of the inclusion of a planetoid upon the early globe as herein described may be better shown by a diagram. Let fig. 22 A be the metallic core of the planetoid; $b\,b'$ the surrounding tertiary matter; $c\,c'$ the surface of the globe; $d\,d'\,d''$ the mass in section of the tertiary matter finally resting above the surface of the globe; $t\,t'$ and $t''\,t'''$ will show the lines of greatest mass that will rest above the surface. The tertiary mass in segment of arc $z\,z'$ in a heated viscous state would float round the denser core and finally appear upon the earth at d and d'. The extruded tertiary matter would generally finally rest in gravitation-equilibrium with a part of its mass sunk in the

matrix matter of the globe, and being vitreous it would also tend to run down upon the surface, but in its early extrusion, being at the same time subject to the radiation of heat when projected much beyond the earth's surface, it would partially cool down and finally take a mean gravitation position in relation to the dense metallic core of the earth with considerable projection therefrom.

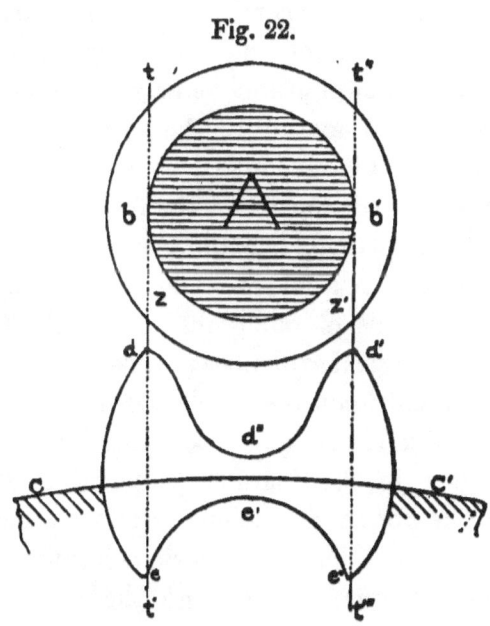

Fig. 22.

232. As the planetoid may be assumed to strike the globe near its equator, it would have more of its matter included upon the equatorial than the polar side of contact. The plutonic mass would therefore tend to extend poleward, diminishing its volume so as to leave the continent due to its projection of pear-shaped outline, with its broadest end towards the equator.

233. The cold planetoid falling towards the earth-centre would leave its upper surface exposed for a period, and this shutting-off the radiation from the highly-heated globe beneath would cause a rapid deposition by condensation of the super-

imposed nebula, so that it would acquire a heavy surface-stratum of new deposition beyond that of the planetoid's mass.

234. In the above we have taken an extreme case. The special condition of the fall of small bodies of meteoric matter could not have produced distinct features upon the earth's surface. If meteoric bodies fell in swarms, the effects would be the same as those just considered for single planetoids, but probably with more dispersion upon the earth's surface. The general detached fall of single small meteorites could only produce a certain amount of incidental inclusion of that which appears to be foreign matter to the system of stratification wherein it fell, which is recognized by geologists.

235. It cannot be assumed that the foregoing conditions of projection of large planetoids could remain permanent, being since subject to long periods of atmospheric denudation and all the after-conditions of our globe, of which we have full knowledge by geological evidences that remain in volcanic and stratified rocks; but upon the whole the initial setting-out of the globe in local land-areas, if it occurred by the inclusion of planetoid matter in the manner proposed, could never be eradicated. The forms of ancient continents have no doubt been modified in time by atmospheric conditions, and by the effects of oceanic currents, through the direction given by solar impulse, causing wear in their projecting parts and deposition of matter in quiescent contiguous water, owing to which the great ocean-basins at the present time approach circular areas of circulation of free oceanic surface, which I have considered in my work on fluids * (fig. 22). Under these conditions the central continental matter may in some cases be undisturbed. The general mass of vitreous matter, localized in great depth of rock by such projections as suggested, could never be entirely distributed by forces known to us as having been active upon them.

* 'Fluids,' 1881, pp. 354, 377.

CHAPTER XII.

Hypothesis of the Formation of the Earth under purely Nebular Conditions.

236. *Earth-formation under Nebular Conditions only.*—This was proposed (§ 204) as the most important factor. The proposition of land-formation from discrete matter given in the previous chapter may be considered as purely incidental, affecting certain land-areas and occurring at indefinite periods. On the other hand, the conditions of nebular condensation to form the earth must have been constant during the radiation of its initial heat at the time that such nebula could form a dense atmosphere around the earth. It will be therefore convenient for distinction and simplicity to consider this nebular condition separately as regards its entire effects, neglecting for the time the conditions which have been already discussed of incidental incorporation of solid planetary matter with the earth. This will not change the discrete conditions of projection of matter before proposed, which must have entered into composition with the nebular conditions now to be considered, as it is in this combination only that we can have the entire evidences of the actual conditions satisfactorily explained.

237. We have already considered many modifying conditions that may have been active during the formation of our earth, particularly in the effects of the state and condition of the sun at different periods. Yet we must assume that there was a general or uniform system of condensation of

the nebula which formed the earth due to radiation of its heat only into space, upon which the varying outward conditions of the state of the sun would be superimposed. Under such purely nebular conditions we will assume for the present that the sun's condensation may be taken as a constant, and that the temperature of the earth's nebula was decreasing at a uniform rate. The more special condition of variation will be considered in the following chapters. The factors of purely nebular conditions will be most conveniently taken under separate headings.

I. Period of dissociation of elements onward to the time of the early condensations which were adapted to finally produce the globe in a liquid form.

II. Period of condensation of volatile metals, association of oxygen and the halogens with metals and metalloids to form the earth's crust.

III. Period of deposition of water.

IV. Period of deposition of snow and of ice formations.

This last period is very important in completing the causes of changes that have passed; although in this we go beyond purely nebular conditions, still the mode of action is continuous. This subject may be better considered in a separate chapter.

These periods, although it is proposed to take them separately, run necessarily the one into the next by intermediate stages, as the separate systems defined depend upon the effects of time which is continuous.

238. I. *Period of Condensation of Highly Refractory Matter.*—At the commencement of the first period suggested above, we may assume that, immediately after the separation of the earth's nebular ring from the sun's nebular system, the heat at this time may have been sufficient for the dissociation, at least so far as a gaseous or vapourous condition, of that which was to become terrestrial matter. At such a time, if the nebular ring were denser in any part or if it

were broken by any disrupting cause, such as the intrusion of a comet, a local density system would be formed in some part, which would then become the nucleus of our earth. The ring of nebula after its detachment from the sun is assumed to be in equilibrium in its mean orbit, so that all parts of its matter would drift slowly towards any denser parts by consecutive attraction of its near parts. In such a system if the nucleus was formed by heavier heated matter than the condensation of the exterior nebulous matter, this would drift thereto and approach the centre in convection currents. These currents would be constantly active in proportion to the difference between the density of the central system and that of the surrounding vapours or gas to cause the specifically heavier vapours, as, for instance, those of the heavier metals, to move through the minor resistances of the lighter gases towards the centre of the system. At the same time light elastic gases would be buoyed up to the extreme outer circumference of the system. This is exactly the case at the present time with the exterior of the chromosphere of our sun, which we find surrounded by hydrogen, the lightest known element, with denser gaseous and vapourous matters placed concentrically beneath *.

239. Immediately following the first condition of an extensive nebulous or gaseous atmosphere surrounding the new globe-system forming from the continued process of radiation of heat, there must necessarily come a period of condensation of the nebulous matter to liquid or solid matter. Whenever this period arrived it would be quite indifferent in what part of the nebulous ring system about its exterior surface the condensation occurred. The condensed matter would, by its superior specific density, immediately proceed to pass through the lighter gaseous parts towards the centre of gravity of the system, even although the system

---

* Lockyer, 'Studies in Spectrum Analysis,' p. 147.

might remain almost entirely nebulous, just as rain passes through the atmosphere. Thus, if we imagine the platinum group of metals to be widely diffused in the nebulous state, these would not only tend to take a central position by gravity, but in all possible condensation through exterior radiation they would flow towards the centre. Particularly in this case, as such matter as platinum would have little affinity to unite with the oxygen or halogens present through which it might pass. Further, this metal would condense at a temperature so high that the more general matter prevailing would remain gaseous. Like conditions to those just explained for platinum would hold also with all other dense refractory elements, causing constant exchanges of relative gravitative positions throughout the long period of condensation until the densest and possibly the least oxidable of the original nebular matter formed a considerable globe.

240. Whatever may have been the composition of the early condensations or primitive liquid matrix which formed the early earth, whether this was metallic or mineral, in the ordinary sense of the term, one thing is certain, that the present crust as we know it is very largely composed of metallic and siliceous oxides. It is therefore reasonable to suppose that oxygen, being lighter in its pure or gaseous state than the average mass of the early nebulous globe, remained in the early highly heated nebula an exterior element, and that compounds or oxides derived therefrom were deposited at a later period than the refractory metals.

Under the above conditions, assuming that the denser, more refractory matters continually condensed about the central system or globe, this condensation might occur even while the surface maintained an almost incandescent temperature; so that possibly at a period when the earth measured not less than within 400 miles of its present diameter, it was still a liquid globe at a white heat comparable in temperature at its surface with that of the lavas which issue at the present time

CONDENSATION OF THE EARTH. 173

from our volcanoes. At this period the globe would necessarily be surrounded by an atmosphere which contained as a part of its earliest gaseous density arrangement nearly all the oxygen, chlorine, and hydrogen of our system, together with all other elements which remain gaseous at the high temperature now considered, except only such part of them as might be condensed or combined near the surface of the globe by the enormous nebulous atmospheric pressure then prevailing. The globe, unless disturbed by collisions with exterior matter, would be at this time an incandescent smooth uniform liquid spheroid, without any projections upon its surface.

241. From the mean density of the earth, 5·6, it is highly probable, as before proposed, that the central matrix is entirely metallic. Under this condition, however distinctly the condensations might form at the earliest stage an increasing density system towards the centre, the tendency of metals to form alloys would materially modify this during the after condensation of the lighter metals. Further, the heat-conducting powers of the metals would tend to keep the globe at a uniform temperature to a depth below the point where matter could experience the effects of radiation.

There is no reason to suppose that these conditions have been materally modified as regards the greater part of the mass of the earth up to the present time, as there could not have been great loss from radiation during the time that the earth was surrounded by a dense nebulous coating which was giving out its heat on continuous condensation, nor after the oxidized coating had condensed to form our present highly non-conducting surface, as this would give out great heat upon its oxidation. Therefore the probability is that the loss of central heat was and is largely operative only in condensing or thickening the surface coating, under conditions which will be discussed further on.

242. At the period when the earth was wholly liquid the outward diameter of the sun or his photosphere probably

extended much beyond the orbit of Venus. At this time the earth was possibly outwardly a self-luminous globe partly obscured and surrounded by a moon-ring or nebulous moon. Possibly also the planet Jupiter was not wholly condensed, but appeared as a large heat-giving nebulous body dispersing a small part of this heat to the earth. Under these conditions the earth's equator by reciprocal radiation-of-heat exchanges with the sun, and in a less degree with the other planets, maintained nearly its initial temperature for the time. The polar regions being open to space would be radiating their initial heat more rapidly than the equatorial regions, thereby representing the area of cooling surface to which condensing nebulous matter near the earth's surface would drift in aërial currents, a condition which, as regards the vapour of water, has remained to a certain extent permanent.

243. II. *Period of Condensation of Volatile Metals, Association of Oxygen and the Halogens with Metals and Metalloids.*—We may now consider the conditions of a period when the sun's photosphere had shrunk to nearly the orbit of Venus, and heat exchanges within the planetary plane had grown much less active; when the polar regions through radiation of their initial heat may have been reduced to a red heat and assumed a viscid consistence. Under these circumstances, by the continuous radiation of heat from the exterior nebulous surface or atmosphere, and the cooling by radiation of initial heat from the surface of the globe, volatile metals would be deposited and oxygen and the halogens would approach more nearly to the earth in its nebulous atmosphere, by which means their association with metallic vapours near the surface would be possible and might even be rapid. At this period, under the great nebular atmospheric pressure then present, the new-formed oxides would fall as rain over the cooler polar regions, where the surface heat could not support the vapourous state; and we may assume that at their great heat they would remain

in a viscous state, and therefore would flow outwards from the cooler polar regions towards the equator to gravitative equilibrium, combination with the central matrix being no longer possible, as in the early conditions with prevalent metallic surface matter.

244. In continuity of what is here taken to be the same period we come upon conditions most important to geology, namely *when it became possible as a natural result of the cooling for solid oxides, when they formed tertiary matter, to rest upon the surface of the globe.* This occurred at first possibly as a kind of oxidized scum, such as we find floating in the ordinary melting of metals exposed to the air, and no doubt only over the colder area of the poles as just stated. The presence of such a scum acting immediately as a non-conductor of heat would shut off a part of the initial radiation from the surface of the central matrix of the earth, which previously kept the gaseous matter above it from condensation at nearly its critical point. This loss of the support of radiated heat from the earth would cause the condensation of vapourous and gaseous matter to appear as dense clouds and finally to precipitate the condensed oxides, of which the clouds were formed, upon the cooler polar areas of the earth. Under these conditions, as soon as a cooler primitive scum was formed, this would be immediately covered by further nebulous deposition of oxidized matter.

245. As the polar regions of the earth cooled down through open radiation the early basic matrix would no longer maintain sufficient surface-heat to retain a liquid surface, and the early siliceous oxides would become by loss of heat stiffly viscous. After this period, the same oxides that fell at great heat in a liquid or viscous state might fall over the cooler areas as a kind of tertiary mineral snow, and in this way cover up the primitive viscous scum, possibly even to very great elevation. At the same time this would depress the supporting liquid and the viscous surface beneath, causing

this surface-matter to move outwards from the poles to equilibrium of gravitation by displacement. In this manner gravity would spread out or carry the newly-formed oxides to great distances from and around the poles.

246. In the above scheme it is assumed that oxygen and other elements of low specific gravity at a high temperature could not reach the earth for combination with denser metallic matter at any early period of condensation. But if we suppose there were certain combinations which would be precipitated at any given temperature from the nebula, even before the complete metallic matrix was formed, their specific gravities would still maintain them as a scum, and the after conditions, as already described, would not be materially modified.

247. Long after the deposition of the more refractory oxides considered above, we should naturally have the deposition of the more volatile chlorides, and in a less degree the fluorides and iodides, and generally we should have the most volatile matter of the solid earth longest in suspension in its primitive nebular atmosphere. At this point it therefore becomes important to consider what bodies in combination may have remained gaseous at a moderate temperature capable of finally contributing to the matter of the earth's crust, which we may assume, from the presence of their elements in the early rocks, were possibly largely diffused previously to the formation of our present atmosphere. Of these gases as most important we may take:—

Aluminium .... Aluminic chloride, Al_2Cl_6.
Iron ... . Ferric chloride, Fe_2Cl_6.

Or, descending to our present atmospheric temperature, we have:—

Silica .... $\begin{cases} \text{Silicic hydride, } SiH_4. \\ \quad \text{,, chloride, } SiC_4. \\ \quad \text{,, fluoride, } SiF_4. \end{cases}$

Carbon .... $\begin{cases} \text{Carbonic oxide, CO.} \\ \text{Carbonic dioxide, } CO_2. \\ \text{Carburetted hydrogen, } CH_4. \end{cases}$

Sulphur .... $\begin{cases} \text{Sulphurous anhydride, } SO_2. \\ \text{Sulphuretted hydrogen, } H_2S. \end{cases}$

248. To which we may add the vapour of water. In some cases one of these gases would decompose the other and cause deposition. There are other compound gases, less important in a geological sense, that would remain constantly present at a somewhat higher temperature. There is an omission in this volatile series of two very prevalent surface materials, *calcium* and *magnesium*, the chlorides of which are gaseous only at high temperatures. But the oxides of these bodies are so extremely light, that assuming they condensed with the heavier siliceous oxides they would fall very gently through a dense resisting medium, such as the nebulous atmosphere at this period is considered to be. Further, from these oxides being very light and refractory in a finely-divided dry state, they would remain upon the surface and drift about in the dense aërial currents then prevailing, and rest finally in positions above the heavier siliceous and denser metallic compounds, which would be for the greater part condensed at the time. Under any conditions, by their specific gravity they would take finally a surface-position upon the primitive scum. The condensation of the above-described compounds with others present, possibly formed a considerable part of the superficial crust covering the earlier deposited oxides; but at this time we commence to have the interference of water condensed in large volumes, which was instrumental in many changes to be considered further on.

249. *Formation of Continents.*—We may now consider the effects of the above-described formations generally upon the earth's crust as influencing the formation of continents, under conditions of nebular condensation only. We may

take it that at the period when oxygen and the halogens entered into association with the prevalent basic metals, their oxides were first deposited over the cooler polar areas upon principles just discussed, that all the surface of the globe, except the polar regions which were most open to radiation, was in a highly heated liquid state, and the whole substratum of the polar area in this state also. Under these conditions, as the oxides condensed about the cooler areas of the poles this lighter oxidized matter, its lower part being in a plastic viscous state loaded up as it might be by condensation at the poles to great height, would have through the influence of its gravity a constant tendency to be pressed outwards from these regions, so as to float upon the surrounding liquid metallic matrix into lower latitudes. At the same time, by natural affinity these oxides would cohere together, more particularly at the base of the mass, where the internal heat and pressure would maintain them in a liquid viscous state. Thus as the compound oxides accumulated in mass they must of necessity have overflown or be thrust outwards from the cooler polar regions from their own internal gravitation-pressure, and in this manner form coherent projections, taking the lines of least resistance in moving upon the liquid or semi-liquid metallic matrix constantly towards the more highly-heated regions of the equator to restore gravitation-equilibrium. This thrusting-out would occur exactly in proportion to the accumulation of the lighter condensed oxidized matter from its increased pressure through constant deposition at the cooler area of the poles.

250. The form which the above-described coherent projections of oxidized matter would take would be that of long tears, precisely the same as that of the boundaries of a melted metal or other cohesive liquid poured out continuously in bulk upon the centre of an extensive heated level slab. This form would, however, be modified to a certain extent by the re-melting of part of its extreme prolongation, particularly at the

thinner bordering edges in approaching the more highly heated, that is the less cooled, equator. These prolongations from the polar regions would therefore take the final forms of pointed tongues, maintaining a wide base upon the polar regions and diminishing gradually as they approached the still almost incandescent tropics, where heat would be partially held from radiation by the large predominant sun.

251. Under the above conditions there would be also present a constant tendency to widen the base of the prolongations above described. Firstly by the extension of the solid surface of the polar regions through continued cooling and the increase of the amount of support to the elevation of oxidized matter through submergence of the lower parts. Secondly, by the deposition of a greater amount of nebulous matter, which would increase proportionally to the surface shading from the influence of the radiation of the earth's initially-heated matrix, so that from both these causes a constantly wider area of base would be added to the floating prolongations as they were pressed outwards from the polar regions, making the entire projections of conical form. At the close of the period above depicted, so far as condensing nebular conditions were concerned, the whole of the new-formed land-surface, derived from the deposited oxides, would consist of pointed areas of projection extending towards and over the equator, proceeding from an elevated base spread over a wide region around the poles. The new land in lower latitudes would appear as a floating mass, possibly of a dull red heat, upon a liquid matrix of much greater density and more highly heated than itself.

252. We may now consider a period when the whole surface of the globe proceeding from the poles began to set to a viscid resisting surface as an effect of the radiation of its initial heat into space, and when the new land-surface in pointed continental areas had become superficially rigid. At such a period the central areas of the new land, being pro-

tected by a non-conducting coating from excess of radiation from the still highly-heated matrix beneath, would possibly retain a semi-viscid condition at a small depth. We still assume that mineral matter is being deposited, although more slowly, from space, especially at the cooler areas of the poles now assumed to be much below red heat. Then, however much mineral matter should be piled up at the poles of the earth, it would be more and more resisted at the borders of the now viscid surface of the matrix surrounding the new continents. Under such conditions possibly the edges of the new-formed continents would be pressed up and contorted by the resistance to further projection and have their borders thrown up much beyond the average surface, such edges being continually cooled and consolidated by their exposure to radiation at the greater heights. New smaller tongues of projection might break out here and there at any points of least resistance for the escape of the internal pressure of the semi-liquid oxides flowing to equilibrium.

253. At a later period, when the floating movement of the continental areas became quite impossible by the resistance offered by a considerable depth of the cooled surface of the matrix and oxidized magma above, and the various matters of the nebulous system were mostly deposited, the more heated interiors of continents then formed where the polar outflow was most continuous, would contract and sink down in cooling, and leave the new land-areas, although of great altitude, of basin-like section.

254. *Distribution of Land-areas.*—To summarize the conditions here proposed: the continents may have been fully delineated before water yet filled the oceans, these appearing as pointed land-systems proceeding equally from each of the polar areas; but if we consider the actual condition we find we have very little land in the southern hemisphere and much in the northern. There must, therefore, have been, even at this early period, some modifying conditions to those now

generally proposed to account for this. What such conditions were has been already suggested upon the precipitation-theory in the last chapter. They might possibly also, perhaps less logically, be suggested upon the nebular theory simply.

255. If at the period when the earth was still an incandescent globe, we assume that radiation of heat from any cause might have been greater in the northern than in the southern polar regions—either from there being cooler space towards the northern pole or from deficiency of heat-radiation from distant celestial bodies, or from excess of winter eccentricity of orbit at the time, or other causes sufficient to chill this pole first,—then the first accumulation of oxidized deposits would immediately shut off radiation of initial heat from the early matrix of the globe into the superimposed nebulous matter and cause its precipitation at this pole only; and as the area thus shaded from internal radiation by the first early deposits would constantly increase in dimensions proportionally to the depositions, the pole first chilled by any cause would become in increasing ratio the greatest area of deposition. In this manner the land-forming oxides might be deposited at the north pole in quite sufficient excess to account for the difference we find at present between the amount of land-surface in the northern and that in the southern hemisphere, or be even sufficient to disturb the position of the centre of gravity of the earth's mean volume. So that at the present time the superficial subaqueous region of the south pole may be of much denser matter than that of the north where the early oxides were most deposited, which condition must continue permanently.

256. If we now take cognizance of the present form and distribution of land-surface and oceanic basins, and make allowance for all reasonable changes, volcanic and other, that the immensity of time must have brought about between the periods already discussed and the present, we may ask, Is the distribution of land entirely consistent with the

theory proposed upon nebular conditions only? This would be very difficult to answer in the affirmative, but we may find that in certain points *there is agreement*, which may possibly be all we may expect after so great a lapse of time, and with other factors of formation that have been already proposed.

257. By the nebular system here suggested, we should have a mass of land at the north pole from early condensations, and a less mass at the south pole from later condensation. Both of these land-areas would possess pointed prolongations extending outwards to or possibly even over the equator. Of the actual land-areas which may be considered as consistent with the principles discussed in this chapter, we may take the prolongations of land in North America with arctic base, the lesser prolongation of Greenland, the prolongation of the Europæo-Asiatic continent to Mocha, that of the peninsula of India within the Ganges, and that of Siam, extending at one time possibly to Sumatra. We now find the most important areas of S. America, Africa, and Australia disconnected from polar areas. But as regards Africa, if we take a wider view, this may be considered as a prolongation of the Europæo-Asiatic area, leaving then only the difficulties of accounting for the division of the narrow Mediterranean and Red Seas. In the same manner, Australia having the broadest base to the south, we may imagine it to have been once connected with the antarctic polar system. At this period its area would extend in pointed form, including and possibly extending beyond New Guinea. The principles discussed leave South America an entirely detached continent, which, considering its theoretical form upon the nebular principles just proposed, would indicate that it must belong to the northern area, but we now find it in the southern. Therefore, this may possibly be better accounted for upon the discrete conditions previously suggested.

258. It is not probable that the early conditions of earth

DISTRIBUTION OF LAND-AREAS. 183

formation could be fully recognized, as very material changes must have occurred since. The anomalies to the theories proposed of the detachment of S. America, Africa, and Australia might possibly be met in a certain degree by supposing that these cohesion systems were detached from the area of their polar condensations and drifted when the earth's metallic matrix was still superficially liquid into their present position. This might have been caused by outward pressure from the poles, of matter which condensed at a certain period, but afterwards dissolved in the highly heated waters, deposited in great volume and under great pressure, which will be presently discussed. Taking all these matters into consideration, however, it is not probable that nebular conditions alone ruled, as there were no doubt also discrete condensations about and without the nebular zone or ring, which in time, through crossing orbits with the earth, came into collision with it and materially modified the land-areas, possibly in the manner discussed in the last chapter.

259. Following the above-described hypothetical conditions, under which *the basic superficial system of the globe* may have been formed, we have for the consideration of the present continental forms to allow for the influence of forces acting onward to the present period, to which we must attribute great modifications. Among such influences we have erosion and after deposition on coasts, vulcanicity in its widest extent, and the wear of oceanic currents, forming altogether constant elements of modification with others to be discussed.

260. Whatever may have been the special or local conditions of early continental formation, it is very probable that at the close of the long period of deposition of oxide and halogen compounds, or what we may term the dry period of deposition, the land-areas were generally clearly defined for all future time. If these oxides as non-conductors stopped the radiation of central heat and at the same time radiated

their initial heat from the surface, the deposits would be piled up over such areas, and by this loading press down the lower surface of the land to gravitation-equilibrium, so that the land of the earth may be represented as consisting of masses of oxidized matter floating upon a lower denser liquid metallic substratum.

261. From the immense volume of oxides known to exist upon the surface of the globe, by the mode of precipitation here proposed the land-areas may have attained great height, possibly locally of ten to twenty miles. These areas, obstructive to radiation of internal heat, would advance from the poles, so that afterwards deposition of oxides or halogens over such land-areas would become general, and upon points of elevation as great as over polar areas, in the same way that snow falls on mountains at the present time. In such new-formed depositions, generally pressing from their centres outwards to gravitation-equilibrium, acting with deposition of water, we have possibly the entire factors of the great gneiss-forming age, vestiges of which remain intact with all the great displacements and contortions impressed upon it which are still evident at the present time. This will be discussed further on. The depositions about the tropical land-regions which may have been produced by causes discussed in the last chapter, would act in conjunction with the condensations here considered and locally magnify their effects.

262. III. *Period of Formation and Deposition of Water.*—It has been calculated that the whole of the water upon the globe, if equally distributed, would be about two miles in depth. If this, at the temperature then prevailing, formed an atmosphere about the globe it would produce a pressure of about 400 atmospheres, or 7000 lbs. per square foot of surface. It is unnecessary to say that this could scarcely occur at a time when the earth's surface was even at a red heat, for at this pressure water would possibly remain liquid even at a white heat. Further, by the classical experiment of Cagniard de

Latour*, water at a temperature of 412° C. was found to dissolve glass. Further experiments have shown that siliceous rocks could not only be dissolved in water by heat and pressure in the presence of alkalies, but be recomposed to form quartz, felspar, and other minerals†. At the heat and pressure then present, water would therefore combine with mineral matter, which would be crystallized and become solid rock as the temperature afterwards diminished. We must therefore suppose no distinct definable line for the formation of water on the globe, but merely make a broad assumption that it probably formed after the deposition of the mass of the most refractory oxides which could remain gaseous at a very high temperature only. Further, there is no doubt that the energetic action of the early highly-heated water would be greatly increased by the presence of corrosive mineral acids within it, which would be in a vapourous state at the same temperature.

263. We may imagine at the first great deposition of water which occurred at the cooler area of the polar regions, that by its great solvent power when at great pressure it would at once reduce and dissolve many of the oxides, and more particularly the siliceous ones. That the streams produced would channel out and disintegrate the solid elevated siliceous deposits then covering the polar areas, forming great rivers, and finally, by the heavy constant highly-heated rains, divide the land up into detached parts or islands, carrying the dissolved matter into the deep equatorial oceanic basins to gravitation-equilibrium. That as the tropical areas by the conditions of mutual heat-exchanges would be much more highly heated than the poles, this water would again, for a large part, evaporate over the tropics and leave behind the mineral matters crystallized at the bottom of the ocean. In this

---

* 'Annales de Chimie,' II. xxi. & xxii.
† Geol. Soc. Trans. xv. p. 103.

manner the ocean bottoms would be formed from the débris of the early polar condensations, and the polar areas would become channelled and in time become largely oceanic by this constant denudation and chemical action of the constant drenching of the highly heated polar rains.

264. If we may assume that the process suggested above continued until the sea reached to within half a mile or so of its present level, the condensation of vapour might then have become possible over the now much cooled lower latitudes, and the elevated tropical lands possibly began to receive a copious rainfall. It is throughout this period until the ocean became of nearly its present level that we may look for the greatest aqueous deposits upon all lands in inland seas and shores together with a large deposit over the bottoms of the oceans. The deposits as they were washed from the coasts and inland would be left by gravity piled up against the borders of the continents, and, generally by the action of currents carrying mineral matter, irregularities and depressions upon and about the continents would be filled up or smoothed off *.

265. Of the gaseous matters previously considered as possibly remaining longest in the atmosphere and incidentally under decomposition forming surface matter (§ 247), the most important are the chlorides, and these are not found largely in the earlier rocks, but in the ocean. The reason for this is probably that they were decomposed in the presence of oxide of sodium; and as sodium does not form a gaseous compound with chlorine at ordinary temperatures, it would be at first precipitated on the earth from the nebulous envelope as one of the later lighter oxides, and this oxide afterwards in the presence of water would decompose the chlorides present, and still remaining in solution would carry the chlorine to the ocean as salt, leaving its oxygen reunited with elements

---

* 'Fluids,' by the Author, p. 373.

present which were not saturated to the highest or most permanent state of oxidation.

266. If we take the termination of the aqueous period to be that at which ice was first possible of formation at the poles, this would be the time of greatest general elevation of the ocean; for although we may suppose that a much greater volume of water was held in vapour in the atmosphere by the heat conditions present, still this would not nearly equal the enormous amount of ice at present elevated at the poles above the oceanic level which in the case assumed would form part of the ocean. Dr. Croll * has estimated that the melting of the antarctic ice-cap alone, considering this as only equal to one mile in thickness, would raise the general oceanic surface 200 feet. A mile is probably much under the truth for the elevation of the interior of this area; but if we add to this the arctic ice above sea-level, and that of all elevated ice-clad regions, the general level of the ocean would be raised possibly more than 400 feet from its present surface, which, added to the present sea-level, might indicate approximately the level of the ocean at a period before ice was formed. Further, it may be presumed that the constant deposition of water must have naturally reduced the amount of land, so that the lower lands of the present surface were entirely submerged.

267. This aqueous period, as will be hereafter considered, possibly occurred in the palæozoic age and also later in the miocene period, when a temperate climate extended to the north of Greenland. The general effect of these aqueous periods upon what was at the time, and what afterwards became, low continental land, was the heavy and almost continuous deposition of the débris of originally deposited matter brought down by constant rainfall to the shallow coasts and inland seas. This deposit upon principles suggested, although

* 'Climate and Time,' p. 388.

less in proportion as it was distant from land surface, must have been greatest towards the polar regions, partially filling the oceans in these regions and generally diminishing towards the equator. Therefore it is improbable that such enormous deposits as the northern silurians will be found in the tropical regions, their representatives being much thinner strata derived from the débris only of the elevated continental lands, produced entirely by causes already discussed.

CHAPTER XIII.

CONDITIONS OF THE COOLING EARTH DUE TO FORMATION OF
ICE AT THE POLES.

268. *Pre-glacial and Glacial Periods.*—We have now to consider the fourth period (§ 237), when the residual nebular matter surrounding the earth consisted, as at present, of air, water, and carbonic anhydride. During this period, which extends to the present, the earth-surface, from the effects of the sun acting upon it, may be conceived to represent a heat-engine in which the tropical regions are evaporating the surface-water which is being simultaneously condensed about the poles. Under these conditions the probable geological effects of accumulation of snow about the poles will be now estimated.

269. At what period ice could first accumulate about the poles would depend upon the amount of secular cooling of the earth's surface and the amount of heat received from the sun. Difference in the effective amount of sun-heat might be brought about either by diminution of his volume or by clouding effects which may have occurred through condensation at any critical point of temperature of the solar nebula (§ 95). The diminution of the sun's disc would be constant and regular: the clouding effects would be exceptional and depend upon the chemical constitution of the nebula at the time. These exceptional conditions due to clouding will be considered hereafter in another chapter; the regular or symmetrical condition due to the diminution of the sun's

disc will now be suggested. These symmetrical conditions were discussed by me in a paper read before the Geologists' Association, March 2nd, 1882, and are now reproduced with some slight additions *.

270. Ice, of which we have the greatest mass at present, under the uniform condition of condensation of the sun, probably formed when his apparent disc was not more than ten times his present diameter and when the mean surface-temperature of the earth did not at most exceed its present tropical temperature. At this period we had a much more aqueous atmosphere than at present, as it is the law that the quantity of vapour in the atmosphere when this is saturated increases in geometrical progression as the temperature increases in arithmetical. We had, therefore, probably at this period, considering the relative areas of the globe under tropical and temperate temperatures, a mean of about ten times the present amount of aqueous vapour in the entire atmosphere. At this relatively warm period as compared to the present, we have only to imagine that a polar area became sufficiently cooled through excess of radiation of initial heat from the earth for condensation of water to occur in the solid form of snow, and we have then the certainty of a continuous copious fall of snow over the same cold regions during the winter in the place of a former rainfall.

271. If we take the process of the cooling of the earth as being quite gradual, omitting variations in the heat-giving power of the sun which may have been brought about under conditions already considered and other effects to be discussed in following chapters, we may suppose that the cooling of the earth sufficient for the deposition and formation of ice would at an early period produce very little geological change, if we omit consideration of all effects upon animal and vegetable life, which would be materially affected by frost. The

* 'Nature,' March 29, 1883; Proc. Geol. Assoc. vol. viii. p. 89.

ice formed at this early period in the winter would be dissipated in the summer, uncovering the land-surface and leaving it at about the same level, except for a small amount of denudation.

272. After the early period defined above, the immediate effect of the further cooling of the earth from any cause, astronomical or secular, would be the greater deposition of snow in high latitudes, which as it constantly accumulated in mass would slowly bring about the proportionate lowering of the oceanic surface upon the entire globe from abstraction of the surface-water. Under these conditions the littoral areas formerly submerged in shallow water would be gradually but slowly uncovered, until in the course of time the present extent of land-surface appeared.

273. If we assume upon the symmetrical conditions proposed, that the entirely aqueous epoch closed with the miocene period, when ice of the present system, due to decrement of the sun's volume and other causes, probably began to cover the poles in winter and melt in the summer, this would evidently be the period of the greatest extent of oceanic surface, for not only would the waters of the ocean be distributed over its surface to gravitative equilibrium, but the land would have become largely levelled down by the heavy rains of the earlier period, when the atmosphere was more highly charged with vapour. When the elevation of ice at the poles after this period slowly abstracted a much greater part of the waters of the ocean, much of the shallow muddy shores must have become soil adapted to vegetable growth over the temperate regions.

274. The continuity of the oceanic depression upon the conditions just stated and contemporary circumpolar elevation by deposition of snow, as it changed the extent of land-areas, must have affected the land-resistances to the direction of the projection of oceanic currents, and with them the superimposed air-currents, and have caused local variations of temperature

in polar and temperate regions by the direction given to these currents due to heat and expansion of air and water from the diurnal impulse of the sun *.

275. The elevation of snow to great height by condensation at the poles would by its pressure upon the yielding mass of the globe, whose equilibrium could only exist in a form consistent with its gravitation and rotation as a spheroid of revolution, cause an excess of pressure upon polar areas, which would react upon the lower viscous superficial strata of the earth and cause extrusion of viscous rocks to the surface or the elevation of certain areas of the surface which offered the least resistance to the imposed internal pressure until approximate equilibrium of the rotative gravitation system of the globe was restored. This elevation would be brought about by the extrusion of felsitic or basaltic rocks to the surface, or slowly in the upheaval of extensive land-areas, or more violently in earthquakes and volcanoes, according to the state of resistance due to the density, flexibility, or previous faulting of the more or less yielding surface-rocks.

276. It must be impressed that the necessity for the maintenance of land-areas depends upon the *elevation of rocks* by plutonic forces. Degradation by atmospheric forces and tidal action is constant, so that unless the elevation is in excess of the depression in the ratio of this constant degradation, land-areas must constantly diminish by levelling-down, and, judging from the quantity of sediment known to remain at present in stratified rocks, all land-areas must therefore have been wasted away ages ago.

277. *The Distribution of Ice at the Poles.*—Under the conditions discussed § 245, the North Pole would be the first to cool down for the deposition of oxides, so that deeper surface-rocks would form at first at this pole. This deposition being of a large mass of matter would disturb the centre of gravity

---

* 'Fluids,' p. 391; also Brit. Assoc. Rep. 1884, p. 723.

of the globe, bringing it to a more northern position; the effect of which would be that the surface-rocks of the South Pole having the dense metallic matrix matter left nearer the surface would represent altogether a denser mean gravitation surface. The waters of the ocean being free on the surface would flow towards the denser system of the South Pole in equation with the density of lower matrix matter to restore general equilibrium. At a later period, when both poles were frozen, the aqueous evaporation-surface would be greater around the South Pole, as it is at present; and, as the cold is assumed to be sufficient during this period at the South Pole to cause the deposition of snow, it would be greater in proportion to the surrounding area of evaporation. So that the South Pole would gradually acquire and henceforth possess the greatest accumulation of ice. Under these conditions, as regards deposition of aqueous vapour, the phenomena of excess of deposition of matter would be exactly reversed from the earlier state when the North Pole received the greater deposition of nebular matter, as before discussed.

278. The effects of the great accumulation of ice at the poles may possibly be treated most demonstrably by consideration of the present conditions, which represent the terminal extreme effects of the present system of ice-formation, so far as the uniform cooling of the earth has advanced. At the same time we must recognize that the long interval, between the earliest and latest period of ice-formation, must have effected a great many changes upon the earth's surface, as the resistance of the crust has constantly become greater against deflection of its surface, by cooling through general distribution of internal pressures, due to the local compression of ice at the poles. Some suggestions with regard to this point will be made further on.

279. *Present Conditions brought about by Elevation of Ice.*—With this general discussion of active conditions present, we may now proceed to discuss the theory proposed upon these

premises of the effects of the action of elevated masses of ice upon certain portions of the earth's crust. In the first place there can be little doubt that there is at present a great accumulation of ice at the poles of the earth. In the southern ocean this forms a wall about the Antarctic Circle, according to Sir James Ross, of seldom less than 200 feet above the sea-level, where icebergs are constantly detached *. In certain districts these are evidently of much greater height, as we find large icebergs floating with the upper surface said to be as much as 600 feet above sea-level, indicating a submergence of probably five times this depth, or 3000 feet. As these icebergs are detached from the front of the coast, it is quite clear that the ice must flow down from the interior, as in the ordinary glaciers; therefore there must be heavy deposition of snow accumulating at the back of them. Mr. W. Hopkins has calculated that ice will just move downwards at *one degree* of inclination. In taking the inclination of ice over some mountain-valleys in the Grindelwald glacier in Switzerland, which is a very flat glacier for about a mile, I found that the mean for the lower parts of this glacier was not less than *two degrees*. The southern ice-cap includes an area approximately equal to the entire Antarctic Circle, that is of about 700,000 square miles. Taking Mr. W. Hopkins's estimate, *one degree* of elevation (as pointed out by Dr. Croll †), makes the altitude of solid ice about 24 miles in thickness over the southern pole. Such a thickness, assuming the ice by compression to take nearly the solidity of surface-water, would represent a potential force upon the earth's crust of say 3950 tons per square foot about this pole, or taking an area of 10,000 square miles of surface around the South Pole, a pressure upon the crust in this region would be maintained of over 3900 tons per square foot.

---

* 'Voyage to the Southern Seas,' Sir James C. Ross, vol. i. p. 219.
† 'Climate and Time,' p. 375.

280. We may consider further that it is scarcely possible to suppose the ice a floating mass wholly breaking away at the coasts or that it rests upon a *level plane*, the probability being that the surface is extremely mountainous or irregular inland near the terminal glaciers; in this case we must allow for greater friction on the motive plane, consequently for greater depths of ice before slipping can occur. If 24 miles of ice be assumed to continue in a static condition at the South Pole, above the symmetrical earth considered as a spheroid of revolution, it appears to be highly improbable that the crust, supported upon a liquid matrix, could resist this excess of pressure without deflection; and even if it should do so, the accumulation of ice still remains a constant factor until the resistance is overcome. We must further consider that in Mr. W. Hopkins's experiment the free surface of the ice was only a short distance from the artificial inclination measured *. It is possible that in the case of ice hundreds of miles inland, in every way supported by surrounding ice, grounded on a frictional or even possibly a surrounding mountainous plane and at a much lower degree of temperature than in these experiments, downward movements would not be possible at one degree of surface-inclination. Under such conditions ice might be permanently retained, if the earth were sufficiently rigid, possibly at two or three degrees of inclination, as it is in our inland glaciers. With such a local pressure upon a yielding sphere, which we assume the earth to be, we can scarcely imagine the possibility of its resistance.

Further, ice in cooling increases in density, and we can form no exact conception of its rigidity at $-100°$ Centigrade, as it probably exists at this pole. Forbes's observations showed that ice moved downwards in glaciers with velocity somewhat proportional to its temperature †. Accepting all

* Phil. Mag. 1845, vol. xxvi.
† Forbes, 'Norway and its Glaciers,' 1853, p. 234.

the conditions as active upon a deflectible globe, to cause reactions upon its crust, then the elevation of land, earthquakes, and volcanoes could be easily explained. Similar conditions to those defined for the South Pole would hold at the North Pole, although at the present time less effectively.

281. Unfortunately the poles of the earth cannot be reached for exact evidence of the above-assumed conditions of accumulation of ice, but we happen to have in Greenland a similar state active in a less degree and on a relatively small scale. Here the inland ice being elevated above the snow-line, vapour-currents are constantly condensing to snow, which as constantly accumulates. So that the greater part of Greenland is at the present time a complete glacier-mountain of possibly 7000 to 8000 feet of interior elevation. In the southern part we have at the present time land-surface, and here the coast is now known to be sinking for the space of 600 miles *. Exact measurements have not been taken of the rate of sinking; but ancient buildings upon the rock-islands are said to be sinking beneath the ocean-surface, so that experience has taught the native Greenlander not to build his hut near the water's edge †.

282. It is very possible that the rate of sinking is nearly proportional to the increase of weight of snow annually piled up inland. Such pressure as may be produced in Greenland, situated as suggested over a viscous system of matter, will act hydrostatically and be felt elsewhere possibly by elevation in Iceland or Scandinavia; but as the pressures will combine with the general system of the polar pressures in acting upon the heated magma beneath, where this particular pressure is most reactive at present, that is where the crust is least resistant, it is impossible to discover except by observation.

---

* 'Nature,' Sept. 20th, 1883, vol. xxviii. p. 488.
† Lyell's 'Principles of Geology,' vol. ii. p. 196.

283. It is desirable perhaps to prevent misunderstanding to make a slight detour with regard to the sinking of Greenland, pointing out, what is quite evident, that all sinking of land cannot be attributed directly to loading such land with ice, as all sinking surfaces are not so loaded. All that can be asserted is that accumulations of ice, where sufficient, will cause such sinking. The sinking of a district may occur through tipping of an area of surface, possibly through local resistance to polar pressures, which act in the horizontal direction if this offers less resistance than the vertical. Such, for instance, as that of the Runn of Cutch, adjacent to the delta of the Indus, where in 1819 a considerable area of land sank in one district and simultaneously arose in another. Such tipping may also be noticed in the neighbourhood of volcanic islands where there is contiguous local elevation. Possibly also in some cases it may be produced by the loading of the interiors of continents with sanddrifts produced by prevailing dry winds. The whole of the instances of depression are small and local, and do not in any way compare with the numerous great and nearly constant elevations—such as that of the western coast of South America of from thirty to three hundred feet, for a distance of 1180 miles along the coast and for an unknown distance inland *, or that of extensive land-areas in Scandinavia,— which I assume are due to the reaction of accumulation of ice at the poles upon a deflectible viscous substratum of rocks, and consider to be a necessary condition, in the present order of things, to maintain land-areas against the constant degradation from atmospheric causes.

284. *Where Ice-pressures will be most Active.*—To return to the accumulation of ice at the poles as affecting by hydrostatic reaction the assumed semi-liquid interior, and thereby causing the elevation of land in other parts of the

* Chas. Darwin, 'Geological Observation,' p. 209.

globe, we may be assured in the first place that such elevation will be subject to two important conditions of the crust of the earth :—1. The crust may be nearly uniformly rigid ; 2. The crust may be fractured in parts, or possess lines of weakness. We will take the first condition.

285. If the crust of the earth is nearly uniformly rigid throughout its exposed parts, it will be still evident that if there is an ice-cap of some miles in thickness covering both the poles of the earth, this ice-cap will materially add to the rigidity of the parts that it covers ; for we know it is the property of ice by regelation to heal any possible fracture or strain that may occur from any cause, and thereby constantly to present a very rigid mass, particularly if it is retained in a close area or supported by irregular or rigid surface matter, as before stated. In this respect it possesses a property of rigidity not shared by the tertiary matter of the earth, as this remains faulted after it has been once fractured by an internal strain.

286. If we imagine the ice crust to be maintained at its surface at an elevation equal to one degree over the Antarctic Circle, that is 24 miles in thickness at the South Pole, and to be possibly of great thickness at the North Pole, we may then suppose that the land-surface covered by this coating will be highly indeflectible. Under these conditions the greatest effects of the opposing pressures of the poles would fall more nearly upon the tropical regions, where there is the greatest surface curvature, and each meridian would represent as it were a bent bow under the excess of polar pressure; therefore in the tropics there would probably occur the greatest plutonic or volcanic elevation. Now as we know also that the tropical area, from rainfall, is the area of most rapid denudation, and consequently of more active thinning of the crust, we ought also to find evidence of this being the area of greatest volcanic action, and this upon the whole is fairly consistent with observation. The following

diagram, fig. 23, will give details of the conditions proposed. Let N and S be the poles covered by an ice-cap, E and W the equator, T T', T'' T''' the tropical regions.

Fig. 23.

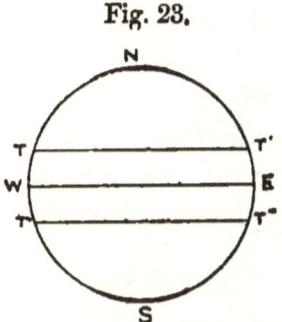

We should then expect the surface of the earth or the outflow of volcanic lavas from internal pressure to rise to the greatest height at T to T'', T' to T''', although this would in all cases partly depend upon the viscosity and other conditions of friction within the strata through which the internal pressures must pass from the region of polar pressures.

287. Taking the above conditions, there are many matters which at once strike one as relative. Thus the mass of ice at the poles, placed originally by vapour forces out of equilibrium with the earth's symmetrical gravitation system as a spheroid of rotation, exerts a force upon the solid crust of the globe, which it overcomes in proportion as it is insufficiently rigid for resistance. In this manner the resistance becomes distributed, so that it is not only from internal or hydrostatic pressures, but by direct horizontal pressure upon the crust that we may have certain facilities of upheaval and deflection of the surface rocks.

288. Further, if we assume a line of expansion over the tropical area, which the elevation of lower or interior matter of the earth and constant denudation really indicates, this would produce also perpendicular strains in the rigid materials of the earth's crust, which would follow the meridians, par-

ticularly in the lines of original polar extension of the lighter elements of matter (§ 250). Thus we should have brought about the conditions of lines of weakness where the pressures at the poles would not act in a direction to close them by any form of crumpling action, as they might be assumed to do in the latitudinal lines. This principle may be shown experimentally by the fracture of an india-rubber ball or a bladder filled with water or air by opposite pressures towards the centre through one diameter, as shown in fig. 24, wherein a fracture is produced from n to s.

Fig. 24.

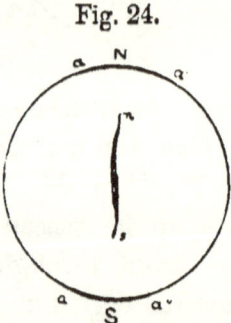

From these causes it is probable that although earthquake regions exist around the globe as an effect of internal pressure, we have the greatest volcanic effects in the tropics, other conditions being equal, which may occur in lines at nearly right angles to the Equator, and continue poleward over land-areas as in the South American range.

289. The above cause of fracture, and the lines of weakness pointed out, may generally rule the positions of volcanic elevations. But another cause of weakness or faulting may be suggested, which is particularly relative to the distribution of ice—that is, that in the immediate vicinity of great accumulations of ice there will be great compression locally upon the immediate surface strata, and this reacting with less friction than at a distant part, may cause faulting near the area of pressure to a considerable depth, as at points

a, a', a'', a''', fig. 24. This may possibly account for volcanoes in the Antarctic and near Arctic regions of Erebus, Terror, and Skaptar-Jökull.

290. *Presence of Water or Steam common to Volcanic Eruptions.* — If ice-pressure were present at the poles, bearing down the surface rocks upon the lower viscous matrix and causing the displacement of the lower viscous stratum of tertiary rocks, the originally cooler surface rocks would necessarily come finally into contact with the central highly heated metallic core. We may imagine that when this occurred the surface rocks as they were pressed downwards by superimposed weight of ice would be melted at their lower surfaces, and become themselves heated viscous rocks. These would be again displaced laterally into the general viscous mass by the superimposed pressure, until a thin stratum only of cool mineral matter would remain as a permanent non-conductor between the ice and the heated liquid metallic core beneath. The metallic core would also in time be chilled to a rigid coating by the constant arrival of cooler tertiary matter abstracting its heat. The weight of ice at the South Pole would constantly increase, particularly in antarctic winter, and this must continually react upon the surrounding viscous matrix beneath, and the surface of the slightly chilled metallic core, just as the former surface rocks had previously acted upon it, as there are no possible convection heat currents in the ice. In this case the former rock-pressure upon the metallic matrix would be replaced by a water-pressure acting through the thin stratum of rocks, which must necessarily continue to act as a non-conductor between the metallic matrix and the ice. Now as we assume that water as it was liquefied from the ice at the surface of the lower slowly conducting rock would receive more pressure above than resistance laterally, it would underflow outwards from the sub-polar area and permit more ice to come in contact with the surface, which would be sufficiently

heated for its liquefaction, so that the process of liquefaction and underflow would be continuous for the dispensation of polar pressures. Further, as the water would underflow from near the surface of the metallic core in an approximately horizontal direction, and at the same time be released from a part of the superimposed pressure in passing beyond the polar area, its tendency would be to flow upwards through liquid rocks nearer to the surface, which would be of less thickness further from the poles. Therefore, convection currents would be impossible backward to the earlier position of the melted rocks. In this manner the underflowing water might become highly heated by the lower semi-liquid rock through which it flowed. This process may be shown by a diagram, fig. 25.

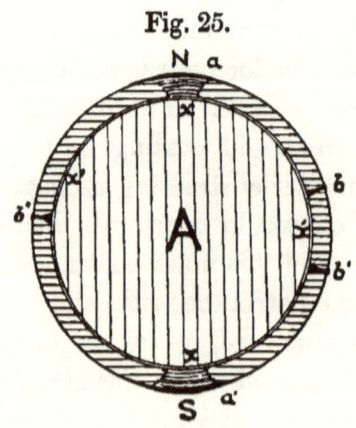

Fig. 25.

291. Let A be the highly heated metallic core of the globe, the mineral coating of the lower part of which is shown at $x\,x$, $x'\,x''$; Na, Sa' the ice-caps. Then by the pressure upon the lower surfaces at x and x, the lower liquid heated matter unable to resist the pressures would be driven to equilibrium of symmetry with the earth's spheroidal form by underflowing in the liquid plane $x\,x$ toward $x'\,x'$. Further, if the ice which formed the caps Na, Sa' by continuity of pressure penetrated to the surface of a non-conducting

WATER IN VOLCANIC ERUPTIONS. 203

stratum at *x x*, where matter was at a white heat within a short distance from the polar region, it would dissolve and carry with it mineral matter from the lower surface rocks, which would be extruded at the first position upon the globe where it could overcome the resistance of surface rocks to establish static equilibrium.

Fig. 26.

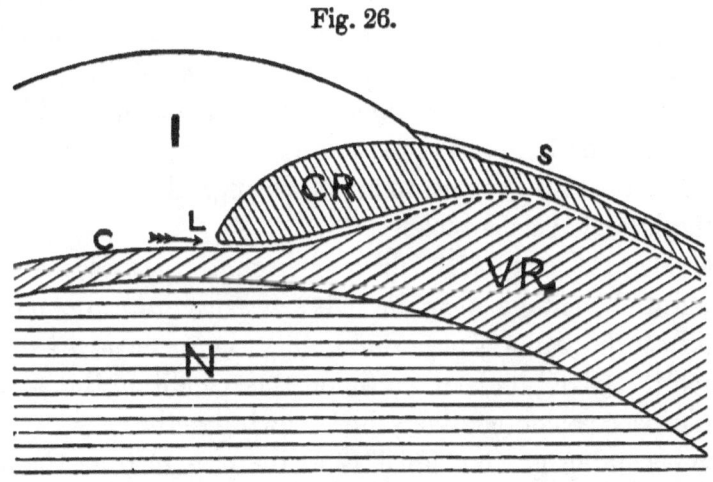

292. Fig. 26 shows details diagrammatically of the manner in which water dissolving mineral matter could underflow the central surface-rocks of the globe. I represents the southern ice-cap. N the dense metallic nucleus-matter. VR the lower heated viscous rocks resting upon the nucleus-matter. CR the present chilled viscous matter forming the surface rocks. C the chilled surface of the lower viscous matter which serves as a non-conductor between the heated nucleus-matter and the ice. L the position where the ice-pressure becomes sufficient to force the lower water beneath it into the viscous matter. This occurs at a lateral position where there is the least resistance to the hydrostatic pressure. The water is forced to form a channel through the viscous rocks into which the remaining water below the ice-cap flows. As the water flows from the polar sub-basin it becomes heated and

expands possibly to double its former volume and dissolves mineral matter from the channel it has formed. The aqueous-mineral matter being of less specific gravity than the viscous rocks, flows towards the surface of the globe, where it may be projected as volcanic matter, or by its hydraulic pressure float up the surface rocks.

The action of release of polar pressures will be generally paroxysmal, as the channel once opened by the pressure will continue flowing through backward-pressure until the liquefied ice beneath the ice-cap is exhausted. The channel will then close until the hydraulic pressure again overcomes the resistance and causes it again to break through the chilled surface of the lateral viscous rocks.

A channel originally driven through the viscous rocks would find ventage near the point of polar pressure, but the water at the same time would chill the surface-rocks from which it derived its heat. Under this condition the ventage would become consecutively lower owing to the surface resistance becoming greater, until, as at the present time, the mass of chilled matters shown in our diagram at CR prevent the outflow or projection of the aqueous-mineral matter until it reaches a great distance from the pole.

The water forced to form a channel in the lower viscous matter, in becoming heated after a certain length of flow, would react upon the inflowing current through its expansion as a resistance. In this manner its injection would become intermittent. I arranged several experiments to show this and found one that would do so very simply. Making the stem of an ordinary clay tobacco-pipe red hot, and plunging it into water, the water enters the tube of the pipe intermittently, and is expanded and projected through the pipe into the air in separate spirts.

The points of departure for the underflow channels, projected by sub-glacial pressures from the Antarctic pole, must occur at the points of greatest extension of land where the subterranean heat can be best conserved by the more conducting nature of the surface rocks. Two such channels when they are open leave the Antarctic circle at about 60° W. long. for the South-American range, and about 140° E. long. for the Sunda Isles. Where the land pressure is not great near the polar pressure this may continue to be a point of least resistance where a fault is once opened, as in Mounts Erebus and Terror in the great Antarctic bay, 170° E. long., and cause eruption at this point.

In this manner a volcano would extrude viscous mineral matter only at its early stages, the backward pressure being upon liquid rock only; but assuming that the volcanic crater continued to represent the point of least resistance to the underflow of the system, the continuity of the volcanic outflow would open out a channel for the following extrusion of water from the pole, which would carry with it dissolved mineral matter, or what becomes, when reduced to atmospheric pressure, volcanic dust and steam. This might finally appear at such distant points as b, b', b''. To find evidence of the above stated hypothesis, which I had previously introduced in a paper read at the Geologists' Association, March 1883 *, I examined very carefully under the microscope the volcanic dust that was blown from the Krakatoa eruption in August 1883. I found it to consist principally of very thin spheroidal surface plates of volcanic glass (*bubble plates* as I termed them) of only about $\frac{1}{15000}$ inch in thickness, resembling on a small scale broken watch-glasses. The vesicles, of which the bubble plates were the remains, were evidently originally filled with steam from the boiling mineral matter beneath, which expanded by release of atmospheric pressure, until they burst when cooled down high in the atmosphere. In this manner they added the effect of expansion to the polar pressure which propelled them through the neck of the crater, and therefore projected

* 'Nature,' 1883, vol. xxvii. p. 523.

them to great heights in the atmosphere. The dust I examined was swept up from the deck of the bark 'Arabella,' sailing in the Pacific, at 1000 miles east of Krakatoa *.

It is not necessary to discuss the manner in which heated water at high pressure dissolves rocks in detail. Water is proved experimentally at a temperature of 412° C., and at a pressure of 100 atmospheres, to occupy about four times its original volume, in which state it dissolves glass rapidly †. At a higher temperature it dissolves siliceous rocks, so that this could very well form an underflowing current of siliceous matter to the volcano, upon the principles suggested.

293. In all volcanic systems, irrespectively of internal expansion of projected matter, there must be a tendency, upon the principles discussed, for the volcanic outflow to rise to hydrostatic equilibrium with the polar pressure. Thus such constant open volcanoes as Kilauea may represent safety-valves of this pressure; but as this mountain is not so high as some other volcanoes which have extruded lava recently, it is quite clear that upon my theory the friction of the undercurrent of viscous mineral matter must withhold a part of the hydrostatic pressure caused by elevation of ice at the South Pole. This is also evident in the differences of altitude of Mauna Lao and Kilauea, which are near together.

294. We may conclude that if a volcanic vent rises nearly to equilibrium with the distant pressure system, the exposed mass will cool quickly in the atmosphere, and the former point of least resistance may become the point of greatest resistance, and the volcano become permanently extinct, or it may overcome surface resistance nearer the base of the mountain than the original crater. If the volcano does not rise to the point of equilibrium it may become intermittently

---

* Quart. Journ. R. Meteor. Soc. vol. x., July 1884.
† Cagniard de Latour, Ann. de Chimie, sér. 2, xxi. & xxii.

active with others in the same state, each eruption evidently clogging or preventing by the weight of ejected matter in the crater future eruptions for a time in the locality, and making the eruptions thereafter more paroxysmal. If we omit the friction of the system from consideration, then the pressure of the highest vertical column of the chimney of a volcano may be taken to represent the hydrostatic pressure upon the liquid matrix beneath, and from this the height of ice at the South Pole might be estimated. For this, Chimborazo might be taken. But the evidence just quoted of Mauna Lao and Kilauea shows that friction may form a large factor of resistance to the distribution of polar pressures, so that equilibrium by ventage cannot be estimated.

295. It is impossible within the limits intended for this work to discuss popular theories with which the above may at some time come into competition, but it may be well just to mention the two most popular.

Firstly, that volcanic and plutonic phenomena are due to the shrinkage of the earth from cooling, which theory appears to persist in our text-books. This is fully worked out by the late Mr. Mallet in over one hundred pages of the Phil. Trans., 1874-5. But the whole matter is built upon inexact data, and there is such great error in the calculation that this must in time suppress this weak theory*. Observations of stratified rocks give evidences almost universally of vertical separation by open cracks, which indicate quite the reverse of a horizontal surface pressure, such as would be the certain result of an internal contraction from loss of heat, demanded by Mallet's theory to cause the elevation by crumpling up of the surface rocks. The universal open cracks are, on the other hand, the natural result of the effect of elevation by plutonic forces as herein proposed.

Secondly, the theory that volcanic phenomena are pro-

* Appendix B.

duced by the expansive force of steam, originally proposed by Spallanzani in 1788, but best known by its development by Scrope *. This theory could not have been proposed with a better knowledge of physics. The experiments of Cagniard de Latour before mentioned, in which water was made red hot at a temperature of 412° C. with an expansion of only four times its volume in a hard glass tube without bursting, demonstrate that water could not act effectively in the projection of rocks as in volcanic phenomena under a very moderate pressure of superimposed rocks, say one hundred feet of liquid basalt, where the water might exist of a white heat and of not over double its ordinary volume. Of course, if the rocks were dissolved in white-hot water, as they would be at this temperature, the water would expand into steam when the pressure was released by its coming to the earth's surface; but this is very different from the assumption that steam at such pressure is able to cause the elevation of thousands of feet of solid rock.

296. There is a further condition frequently offered to support the steam theory in the assumption, in opposition to hydrostatic laws or the observed conditions of heated rock, that water may percolate rock at a low level from the ocean and be projected at a high level where the superimposed pressure of surface rock is greater, which is evidently impossible. Neither are deep-seated rocks, as we find them in deep mines, porous enough to admit of such percolation even if it were sufficient to account for actual phenomena.

297. The conditions stated in this chapter relate to deposition of ice since the miocene period. There were possibly earlier depositions of ice, the conditions of which will be discussed in the next chapter. But I anticipate that all the greatest effects of polar compression by ice, and therefore of volcanic eruption and plutonic action, followed the miocene

* Scrope on Volcanoes, 1825, p. 17.

period, which, I think, from the greater variation of animal life and its higher development, was a longer period than any geological epoch that preceded it. This I will discuss later. Geologists are generally of opinion that volcanic action has been constant. I am not certain that the evidence is clear on this point. If a volcanic vent is found penetrating the silurian surface rocks, there is no reason why it should not be, in some cases open to observation, a tertiary one, as volcanic forces penetrate all surface rocks. Neither is it necessary that igneous matter should reach the surface, as at a certain pressure it may float up the surface rocks to form mountains without any extrusion ; and if these are afterwards degraded, the volcanic vent will appear as though it issued at the stratum level at which its intrusion occurred. At the same time there may have been exceptional conditions of ice-formation at early periods, which will now be considered.

CHAPTER XIV.

PERIODIC CONDITIONS OF EARTH-FORMATION PRODUCED BY EFFECTS INCIDENTAL TO THE NEBULAR CLOUDING AT INFERIOR PLANETARY FORMATION, AND AT CRITICAL TEMPERATURES OF MATTER SURROUNDING THE SUN.

298. *Condition of the Sun during Earth-formation.*— Taking the earth system to have been abandoned by the sun as a nebulous zone according to the theory of Laplace, the earliest condition of the sun in relation to this zone would be that of a luminous globe of nearly the diameter of the earth's orbit. As the earth-zone commenced to condense and form a planet, the sun's volume would at the same time be also condensing and leaving this zone more free by distance. If there was any inequality of density in the zone-ring, the most probable condition, there would then be a separation at the most attenuated part of this zone, and a further condensation in another part by the continuity of the system of attractions of the denser matter. So that it is probable that the earth condensed into a nebulous globe by concentration before any solid matter was formed at its centre.

After the detachment of a nebular planet-zone, this zone having much greater surface area relatively to volume in comparison with the voluminous sun, and being open to free radiation into space in all parts not directly facing the sun, would condense much more quickly than an equal volume of the sun's nebula.

299. In considering the condensation of the nebulous zone

as a special exterior formation, we may conclude that no condensation could occur unless the radiation of its heat into space exceeded that derived from the condensing sun, although at the same time the nebulous zone must have continuously absorbed the sun's heat falling upon it. Under these conditions the radiation of heat into space from the zone must have been of its initial amount plus that constantly received from the solar centre. If we imagine, what is most probable, that at the early time of condensation by cooling a clouding would be produced in this zone through excess of radiation, then the sun's rays absorbed into the zone would be diffused within it, so that the radiation would take place from the zone in all directions, but more particularly in the exterior parts and those perpendicular to the plane of orbit where it was most open to free space.

300. Under the above-stated conditions we may consider the effects of the formation of a nebulous planet-forming zone upon the amount of radiation of heat and light that the sun would be able to disperse beyond this impediment to an exterior planet, assumed to be fully formed at the time. Then, assuming the earth already formed, and the sun contracting in volume before the period of the formation of an inner planet, as Venus, a nebular band, or what we may term the Venus-zone, would appear across the sun's disc, which would continue to obstruct a large part of his rays, and this would last until the complete formation of Venus as a planet. The like would again occur before the complete formation of Mercury. We have apparently, upon a large scale, nebulæ in this condition *.

After a nebular planet-zone was detached from the sun its future condensation would depend upon contingent circumstances.

In fig. 27 the possible appearance of the large nebulous

---

* 2244 Gen. Cat., Rosse Nebula, p. 90.

sun is represented partially obscured by the Venus-nebular-zone at a certain stage of its condensation.

Fig. 27.

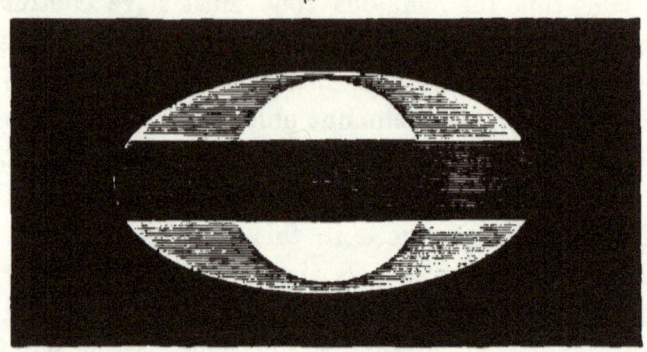

301. If Venus or Mercury, from inequality of distribution of nebular matter after the time when their zone-rings were left by the sun or from any following disturbing cause, as the intrusion of a comet, condensed at first into a nebular globe, as just proposed for the earth; then this globe, after its formation, would in time concentrate to an intensely-heated nucleus at its centre. This nebular system would obstruct the sun's light through cloudiness at its early period of formation; but afterwards for another following period, when it became incandescent, it would present the same light- and heat-giving radiation as that of the nebulous sun to the surface of an exterior planet, or greater in proportion to its visible surface and its state. This new-formed planet, as before stated, would therefore be obstructive to light and heat from the sun if it was nebulous, or auxiliary to it if it was incandescent. In either case this excess or defect of light and heat would occur in periods of the newly-formed planet's synodic revolution in relation to an exterior planet, producing in either case intermittent periods of intensity of radiation of light and heat upon the exterior planet.

302. After the periods of the entire condensation of a newly-formed planet to non-obstructive or non-auxiliary heat-

giving proportions when it could not affect the amount of heat and light dispersed to an outer planet from the sun, the sun would return to its normal luminosity according to its age or state, and become again the only source of heat-energy to the planet.

303. It is not necessary to assume that the sun, when it was an immense nebular globe, gave out in its entirety more heat and light than it does at the present time; it very probably may have given out much less at certain periods. The superficial condensation of nebulous matter upon or about its outer surface would constantly cloud the effects of its intense internal heat, so that when freed from the clouding influence of the formation of a planet-zone, it would present to the earth only the appearance of a magnified image of one of our stellar nebulæ*, or of a large bright cloud.

When the sun's nebular diameter was as great, for instance, as the diameter of the orbit of Venus, if its surface heat had been of the same intensity as at present, condensation of nebulous matter to form the earth would have been impossible.

304. *Distinct Solar Heating Periods.*—Under the conditions just stated it will be seen that the sun's heating effects as regards an exterior planet, as, for instance, the earth, would pass through certain phases or periods from the time of the formation of one interior planet to that of another more interior. Thus with regard to the earth we should have one long period during which the sun would be slowly decreasing in volume, appearing during this time as a large bright nebulous globe or stellar nebula. This condition of the sun would continue, during its condensation, until the time of the disturbing conditions of the clouding effects upon its disc of the commencement of the formation of the Venus-zone of nebula, which would, after a time, appear to entirely encircle the

---

* See 4883 Gen. Cat. Ros. Obs. p. 170, also M 81 and ⩔ 1205 Ursæ Majoris.

nebulous sun as a dark band. The local condensation within the nebular band would ultimately form Venus into a large globular nebulous planet, the clouding-effects of which when moving over the sun's disc would diminish the sun's heat in transit and make it therefore intermittent in intensity with regard to the earth, as just stated, in periods of 584 days. This passing of the nebulous planet over the large nebulous sun at the time of inferior conjunction might possibly at first nearly obscure his light and heat, at other times, exceeding nine tenths of the period of revolution, the nebular sun would appear bright and open.

305. It will be seen on examining the conditions just proposed, which must be incidental to inferior planet-formation as regards the earth, that there were nine somewhat distinct periods of minus and plus solar radiation, therefore of greater or less heat and light radiation, affecting the formation or depositions of matter upon the earth's surface. These may conveniently be defined, to show the effective state of the sun as a radiating body under the entirely nebular conditions proposed during the formation of Venus and Mercury as far as they would affect the earth.

1. *Period of open radiation* of bright nebulous light and heat lasting from the separation of the earth-zone from the sun until the commencement of the condensation of the Venus-zone, during part of which period the earth was a nebulous globe.

2. *Period of nebulous obscurity* of the sun caused by an absorption band across the sun's disc; period of dull nebulous light and heat lasting from the early part of the condensation of the Venus-zone until its formation as a nebulous planet.

3. *Intermittent light and dull periods* of about 584 days, the bright periods much exceeding the dull periods and increasing in brightness from the time of the condensation of Venus to a globular nebulous planet until it became non-

obstructive to solar radiation. Period of great disturbance of local conditions of deposition upon the earth.

4. *Period of intermittent excess of bright nebulous light and heat* caused by the presence of Venus as an incandescent body, particularly near the period of transit, lasting from the time of condensation of Venus to an incandescent liquid or solid planet until it ceased to be in any degree auxiliary to the sun's heat.

5. *Second period of bright open light of the sun,* lasting till the commencement of the formation of the Mercury-zone.

6. *Period of nebulous obscurity by a dense band across the sun's disc,* lasting from the early part of the condensation of the Mercury-zone until the complete formation of Mercury as a nebulous planet.

7. *Second period of intermittent light and dull periods* of about 116 days, the bright periods much exceeding the dull periods, the light increasing with time, lasting from the time of condensation of Mercury to a nebulous globular planet until it became non-obstructive to solar radiation. Period of local disturbance of systematic stratification of disintegrated matter upon the earth.

8. *Second period of auxiliary light and heat,* when Mercury became an incandescent globe, lasting until it ceased to add to the sun's light and heat. The whole period much less active than the corresponding period of Venus condensation.

9. *Third open period of bright light,* lasting from the complete formation of Mercury as a non-auxiliary light-giving planet until the present period of intense solar radiation and for all future time of effective solar energy.

306. *Influence of Inclination of the Orbits of Inferior Planets and Eccentricity.*—It will be seen with regard to the Venus-zone and the second period of dull nebulous light, that owing to the inclination of the plane of the orbit of Venus, 3° 23′, unless the nebular zone had a sectional diameter of about 10 million miles transverse to this plane, it would not

continuously cover the centre of the sun's disc, so that it would appear to shift about from north to south in synodic periods, the entire variation of which would take about 243 years. This would occur until its condensation into a nebular globe.

The same conditions as described above would occur in the sixth period with the Mercury-zone; but in this case we have an inclination of orbit of 7° with great eccentricity, so that the centre of the sun's disc would be covered by a nebular band of 5 millions of miles, and if it were of this diameter transverse to the plane of orbit, its effects upon the diminished solar disc would be nearly the same as that of Venus, but in shorter periods of 7, 13, or 46 years. Under these conditions the dull periods proposed would be in degree intermittent and produce great changes in the atmospheric conditions of the earth.

307. At the formation of a nebulous globe in the early part of the sixth and seventh intermittent periods, it is possible that each of the planets Venus and Mercury, when a nebula, subtended as great an angle at the earth as the nebular sun of these separate periods. In this manner the sun would be obscured at every inferior conjunction and have its disc partly obscured for nearly one tenth of the particular planet's synodic period. This intermittent condition would probably produce greater effects upon the earth's atmosphere, and therefore upon deposition of rocks, than the dull periods, second and sixth, previously considered.

308. It is scarcely possible to realize the enormous effects that would be produced by the obscuration of the sun for a few days only, but very possibly this was only partial. If actual, nearly the whole of the waters of the vapour-laden atmosphere would be precipitated. Terrestrial organic life would be destroyed by the sudden cold. Inland lakes and rivers, and part of the ocean, would be frozen, so that an entire organic change in animal life might follow, this being possibly

EFFECTS OF FORMATION OF INFERIOR PLANETS.   217

afterwards evolved from the preservation of former marine and aquatic species that could retain life only by remaining deep in the ocean. So that, so far as this proposition goes, evolved species may not be found to be by any means uniformly progressive from the highest types throughout geological time, that is, even for the known progressive types.

309. In the above scheme we have considered the effect of the condensations of Venus and Mercury, which were possibly largely induced by the reduction to the critical temperature of the nebular matter which formed the central solar system, § 95. It is probable that this formed only a part of the changes in the amount of heat radiated from the central solar system to the earth. The conditions of critical temperature of the sun have been discussed. The condensation which was capable of detaching a planet-zone would at the same time, as before proposed, obscure a large part of the heat of the central system by metallic cloud. Further, the extensive critical condensations which would form a planet-zone would be exceptional. There were most probably many minor condensations at the critical temperature of certain elements which produced no detachments from the central solar system, but only partially obscured his light and heat. This would certainly occur within the long period following the condensation of Mercury. Possibly such a condensation may have produced our glacial period. Even at the present time the sun's radiation may vary greatly in long periods depending upon the sun's surface density, that is, the amount of condensation at critical temperature tending to cloud the chromosphere. Under any condition the interval period of open solar radiation caused by the simple contraction of the sun's volume when no effects of planet-formation were present would, in all probability, exist for a much longer period than that of any planet-formation. This will be reconsidered. It is only important now to recognize the general effect of the contraction of the sun upon the earth in its more extended action.

310. *General Effects of the large Nebulous Sun upon the Earth's Meteorological Condition.*—Perhaps the most distinctly important condition that would be due to a large nebular sun, free from the disturbing effects of planetary formation, would be that the whole of the earth might be diurnally lighted and heated by the sun, and that this diffusion of light and heat would establish very calm conditions in the atmosphere. The tropics would not be excessively heated and the polar regions would receive at all times direct sun's rays, so that, assuming the body of the earth had cooled below a temperature to affect the evaporation of the oceanic surface, there would not be the excess of evaporation and expansion of air and vapour over the tropics as at present, due to solar radiation, or the great excess of condensation in polar areas, to produce the violent atmospheric currents we have at present in storms and cyclones.

311. At the period when the sun had condensed to the orbit of Venus, the diameter of his disc would exceed 67°, so that it would present a luminous surface of this angle, and the poles, even in mid-winter, would have a segment of the sun shining daily through an arc of 10° versed sine. At the period when the sun had decreased to the dimensions of the orbit of Mercury it would still present an angle of 44°. Therefore, even in mid-winter, there would be at least a daily twilight. Altogether the general conditions of the globe would be very equable as compared with the present, although, of course, the polar area would still be colder, as being more open to free radiation. If, on the other hand, we consider in contrast the conditions that would be active during the shorter intermittent periods 2, 3, 4, 6, 7, and 8, § 305, the atmosphere would at these times be subject to constant disturbance by high winds and heavy rains in intermittent hot and cold periods, which would cause rapid decomposition of rocks, and generally produce intermittent thin planes of varied stratification.

312. As regards depositions of matter upon the earth from

degradation of rocks, these must depend always in amount upon the atmospheric conditions present. We might have a long period during which depositions went on very slowly. A period of light rains and nearly constant sunshine, in which very little matter would be detached from the rocks or brought down to the sea-level, during which time, the conditions of life being constant, there would be very little reason for any change of form, and the strata of universal matter deposited, although uniform, would be very incommensurate with the time. On the other hand, we might have periods of heavy rainfall and frost in high lands, during which time degradation and deposition would be very rapid and the strata formed be very deep for the limited time of formation. With great change of atmospheric conditions there would be great struggles for existence, in which favourable varieties of life-forms only would survive.

So that, on the whole, no depth of geological stratification nor even changes of animal life will indicate, with any exactness, the extent of time of a geological period unless we possess other data for consideration.

CHAPTER XV.

Consideration of Time Elements in the Solar System, particularly for Estimating the Age of the Earth and for its Geological Periods.

313. *Time of Condensation of our Solar System.*—This may possibly be found with some degree of approximation by estimating the rate of contraction of the original nebula upon thermodynamic principles. For this calculation we may take the dimensions of the nebula at any period of its contraction when it occupied a space within the extent of our solar-planetary system until the present time, when it has contracted to the small volume of our sun. In this proposition we again accept the probability that the nebula was originally one of the symmetrical planetary nebulæ of which we possess many instances open to astronomical observation. A probable form being that of one of the nebulæ shown, Plate II. *a, b, c,* or *d.* Such a nebula under the conditions we have assumed throughout this work, must have condensed constantly by radiation of heat from its surface, and at the same time have formed our sun by concentration of matter about the centre. We may form an estimate of the extent of this nebula, at least at a certain period, by the extent of the extreme planetary orbits of our system; and if we knew the rate of radiation at the surface and of concentration of matter at the centre or sun as a heat focus, we could then estimate the time the system may have taken to arrive at its present state.

314. If we omit the consideration of the reaction of condensation of heat in the central nebula or sun, and regard the radiation as a superficial function only, this will dispense the mean energy of the system in proportion to the temperature and extent of surface. If we assume that the heat was equally distributed throughout the volume of the nebula, the surface-heat would be nearly maintained by the superficial contraction and by heat exchanges with the interior of the system. Taken in this manner, a voluminous nebula would condense to less depth superficially in a given time than one of smaller volume.

315. Observations of planetary nebulæ, as, for instance, ⛢ I. 205 or M 81 Ursæ Majoris, Plate II. b and d, show a central condensation of large volume surrounded by a more attenuated medium. Under this condition, which we have assumed throughout this treatise may represent an early stage of our solar system, we have necessarily two distinct systems of condensation—that about the solar centre which condenses its matter from the more attenuated surrounding medium, and that at the surface of the medium itself, which is contracting in space. The final condition of such contractions we assume will form a sun or star, that is, a unit incandescent mass. We may assume that our sun has arrived at nearly its final state. The only representative of the former nebular state being found in the chromosphere. and the largely diffused surrounding matter of which we have evidence in the corona and the zodiacal light.

316. The amount of condensation of the surrounding luminous matter or pneuma into the central solar nebula cannot be defined; but as the condensation must have produced heat at the surface of the interior nebula, which again reacted by radiation, it is not probable that the interior contraction was greater at any time than it is at present. For in the first case, the solar contraction would be blanketed by the surrounding pneuma, and in the present state its heat is

freely dispensed into open space. Therefore we may conveniently consider the present contraction of the sun as a periodical constant without great risk of error, for which we may obtain some data from thermodynamic laws. On the other hand, the contraction of the attenuated medium or pneuma, which formed the limiting volume, as this has practically disappeared from the solar system, must have proceeded at a higher rate. For this calculation we may indulge in certain assumptions which may give approximate results.

317. According to the original calculation of Helmholtz the sun's radius is diminishing at the present time by 1/10,000 part in 2000 years = 1/20,000,000, or 126·32 feet annually. This estimate has since been slightly increased. For an even number we will take 130 feet, which for convenience may be considered as a constant of time-condensation of the solar or nucleus nebula from the early period which we are now considering.

318. As regards the limiting superficial contraction of the pneuma, we will suppose that it condensed in some degree in proportion to its tenuity or distance from the solar centre. At a distant position it may have been rapid; on the sun at present it may have nearly ceased. We will take it that at the period when the solar nebula extended to the orbit of Neptune the mean annual decrease of its radius was, as an extreme condition, possibly equal to two miles. Upon these data time elements may be calculated.

319. For the time of condensation of the solar nebula to the present condition of our sun we will call the radius or mean distance of the orbit of Neptune, in feet, r; the radial contraction of the solar nebula equal to two miles, or 10,560 feet, f; and the constant of central solar contraction $S = 130$ feet. It is then clear that f diminishes, as it is assumed to do, at a uniform rate until it has vanished at the present time, its mean diminution was half this rate. There-

fore the mean total diminution of the nebula for the limit of nebular condensation from the period will be $\frac{f}{2}+S$. This divided into r gives $\frac{r}{\frac{f}{2}+S}=t$, time; which we find to be 2723 millions of years—the suggested time of condensation of the solar nebula from the dimensions of the orbit of Neptune to those of our present sun.

320. Taking the same formula for the earth, calling its orbit-radius or distance from the sun r', or about $\frac{1}{30}$ of r, we have $\frac{r'}{\frac{f}{30}+S}=t'$, which gives t' (earth-time) about 1008 millions of years, or, roughly, 1000 millions of years, for the time of the condensation of the solar nebula from the earth's orbit until the present time.

321. The above calculation may be taken as quite the inferior limit of time; no allowance has been made for the condensation at the solar focus which ultimately formed our present sun, which might increase in volume directly as the increase of gravitation from nearness of the more refractory matter to this centre, possibly in inverse proportion to the distances of all parts of the nebulous matter from the centre. The increasing compression of matter about the centre would more perfectly conserve the heat of the system, surrounded, as it would be, by a denser nebular atmosphere according to Lane's law, § 11, p. 11. Therefore it is not probable that the decrease of volume from the effects of outer surface radiation of the nebula would progress at quite so high a rate as that just stated.

We will now consider the conditions of time-variations of heat and light from the sun during its condensation, from the period when we assume it to have been a nebulous spheroid of the radius of the earth's orbit until the present time.

322. *Distribution of Time upon the Earth throughout the*

varying periods of the Condensation of the Sun and the Inferior Planets.—This may be taken from the calculation just given of 1000 millions of years, making the divisions of time by fixing certain radii of the nebulous sun, to correspond with periods previously defined, for the probable clouding and auxiliary effects upon the sun's exterior radiation during planetary formation. In this proposition we may possibly divide the period of radiant energy into four classes consistent with our table. We may take solar energy to be open—to be obstructed—to be intermittently obstructed and open—or to be increased by the energy of the condensed planet when this was in an incandescent state. The duration of these periods was probably proportional to the rate of condensation. I will take as hypothesis the formation of the nebular zone of Venus to have been obstructive to solar radiation during the sun's contraction for one fourth of the distance between the orbits of the earth and of Venus. That as soon as the sun's nebula was quite free from the Venus-zone, Venus would commence to form a nebulous globe. We will suppose that this nebulous globe remained intermittently obstructive to the sun's rays until the sun had contracted to one eighth the distance between the orbits of Venus and of Mercury. That Venus then gradually contracted and became by incandescence a bright body of possibly not more than six times its present radius. At this time it would be comparable with the sun in density, and probably for a short period much brighter than the nebulous sun, even possibly as bright as our present sun. This state of Venus would slowly cool down, and possibly when the sun had contracted to one fourth the distance between the orbits of Venus and Mercury, Venus had ceased to become in any measurable degree a heat- and light-giving factor to the earth. After the condensation of Venus as a cool planet shrinking to nearly its present dimensions, the sun would remain open until the condensation to form Mercury. Mercury would then go through changes similar to those of Venus in proportion to its mass and its distance from the sun.

323. Taking the above-stated condition that the contraction of the sun's nebular radius remained proportional to the time of condensation, we may construct a table for which we take 1000 millions of years, previously given, as the entire interval between the earth's separation from the solar nebula until the present time.

TABLE.

Conditions of the Nebulous Sun at its radius in Millions of Miles for nine Periods in Millions of Years.

Period.	Radius of sun in million miles.	Difference.	Time in million years.
1. Open	93 to 73	20	114
2. Dull	73 to 67	6	40
3. Open and Dull	67 to 63	4	26
4. Auxiliary	63 to 60	3	14
5. Open	60 to 37	23	197
6. Dull	37 to 33	4	41
7. Open to Dull	33 to 31	2	22
8. Auxiliary	31 to 30	1	14
9. Open	30 to present.	30	532

This table may possibly continue nearly proportional if we consider the specific heat of the solar nebula less and its radiation greater, or *vice versâ*—factors that cannot be exactly estimated.

324. *Period of the Formation of the Nebulous Earth.*—If we take the period from the separation of the earth-zone, as proposed, § 298, and assume this nebular zone was ten millions of miles in cross-section of the annulus, this would give upon condensation, if all its parts moved with equal angular velocity, an excess of rotation to the condensed globe produced therefrom over the revolution of the moon and

the rotation of the earth at the present time (§ 146); but as we presume that this zone would be partially condensed to discrete matter from exterior matter drifting in spiral paths thereto (§ 222), which in falling upon the earth would tend to cause a negative rotation, the cross-section of the zone-ring proposed above may not be too great. This zone after detachment could not maintain its heat to the extent formerly suggested for the surface of the nebulous sun, as we have no large intense centralized source of internal heat comparable with that of the sun within the zone-ring, so that its heat would depend upon its condensation only, and, as this zone would possess much larger surface, relatively to volume, than the sun, it would be open more freely to radiation and contract at certainly quite double the rate of the sun. This would, as before stated, occur principally at its outer surface away from the sun, and at all outward angles transverse to the plane of orbit, as no central heat from the sun could fall upon these parts. In this manner the transverse radiation of heat from the zone would make the total contraction of the zone proportional to its sectional radius. Upon this proposition taking the sun's nebulous contraction at the position of the orbit of the earth as before proposed $= \frac{f}{3} + S$, which will be 482 feet in a year, or at the position of the outer surface of the ring about 500 feet in a year, denoting this by a, the number of miles in the ring-section by b, the number of feet to a mile by c, and dividing by 2 for the contraction of opposite sides of the ring and again by 2 to make the contraction double of the sun's, we have $\frac{bc}{2 \times 2a}$ or $\frac{10,000,000 \times 5280}{2 \times 2 \times 500} = 26,400,000$ years, for the period of contraction of the earth's nebulous zone upon itself to the position of the axis of the ring.

325. In the above-stated proposition the contraction of the

earth's zone-ring upon itself can only be considered theoretically as the *modus operandi*. To give the moon its revolution period and the earth its rotation, we must imagine a separation to have occurred at some part of the nebular zone, under which condition, simultaneously with the contraction of the section of the ring, it was being drawn together by gravitation acting in the linear direction of its circumference so as to form a nebular globe. This would, of course, materially complicate the conditions of the calculation of time by a differentiation difficult to follow; but as the radiation-surface would certainly be constantly lessened during the time of globular condensation, we may prolong the time possible by nearly double the amount of the time suggested, say, to 50 millions of years, for the condensation of our earth from a nebulous zone to a liquid globe at an intense white heat, surrounded as it must necessarily have been by an extensive nebulous atmosphere. This globe would possibly be formed principally of iron with the presence of other highly refractory metals, and become the permanent nucleus of much the greater part of the volume of our present earth, upon which the deposition of oxidized matter may have been superimposed upon principles already discussed. The superficial conditions are represented by the geological periods of the earth, to be considered in the following chapter.

CHAPTER XVI.

GEOLOGICAL PERIODS CORRELATED WITH ASTRONOMICAL PHENOMENA.

326. *Geological Periods.*—Leaving out of consideration any notice of the early paroxysmal school of geology, the eminent geologists, among whom Murchison, Sedgwick, and Miller may be particularly distinguished, who have made attentive study of a single group of ancient rocks, have come to the general conclusion that there have been special periods in the past which have been conducive to the formation of special kinds of rocks, which, with the fossils therein contained, are very distinguishable from other periods. That in and during the different periods for an immensity of time, wide areas of the globe were affected by like conditions from causes unknown. So that if we take, for instance, as the most striking example, the Silurian period of Murchison, this appears to be characterized by the presence of finely-deposited rocks in an entire broad band of unequal thickness surrounding the Northern Hemisphere, containing fossils of a similar succession of faunas, marked by particular zones of genera and species. It is probable also, judging from rock-texture and organic remains, that similar general conditions affected the land-areas of the Southern Hemisphere during the same lengthened period.

A school of thought, founded by the genius of Hutton and developed by Lyell, wherein due recognition, not previously taken, is made of contemporary formations, has arrived at the conclusion that the present agents at work would, if active in the past, account for all former conditions of deposition. The theory hidden in the argument offered appears to be that of the improbability of other changes being active upon the earth than those due to successive astronomical and physical conditions, which are still active, and to the differences of position of land and oceanic areas upon the globe.

327. Quite outside all theoretical induction is the work of the practical or field geologist, who accumulates facts derived from observation irrespective of theory, and makes through observation only a broad distinction between the more ancient rocks, which were in a wide expanse special and uniform per stratum, and the modern rocks such as are at present forming, which, excepting possibly in the deeper part of the ocean, are locally only in very narrow limits similar to each other and vary in character in separate localities in every degree. So that, taking the mean of geological opinion upon the early conditions, there appears to be the extreme probability of the action of conditions in the cosmic system which have entirely passed away. These I propose now to consider as the effects of the astronomical changes already proposed in previous pages which have remained more or less evident in the early stratified rocks.

328. It will not be found practicable in this essay to consider stratification of rocks in detail. This has been well done by many scientific specialists in our advanced elementary treatises on geology, of whom it is only necessary to mention such names as Lyell, Dana, Geikie, Le Conte, and Dawson. It will therefore be necessary to consider in what follows the system of forces already proposed, which were probably active on a large scale as prime movers throughout the distant past under certain special conditions. There must also at the same time have been conditions which were constantly active affecting the deposition of mineral matter upon the earth.

Of this may be mentioned the constant decrement of the sun's volume, the condensations of the solar-nebular matter at critical temperatures, the astronomical constants of variation, of eccentricity of orbit, precession of the equinoxes, and change of obliquity of axis, and possibly also some changes due to the revolution of the magnetic pole which have not yet been recognized.

329. By a general consensus of geological opinion the conclusion is arrived at that the long periods of deposition of surface-rocks which are open to observation can but be representations of detached units of geological time. This is shown most clearly in the great changes of animal remains between one stratum or set of strata and the next, in which the lost period of evolution often appears to be much greater than the long period made evident by the slow deposition of a formation or indeed sometimes of a single stratum. That this should be so may be inferred from the slow but constant action of meteorological forces alone in disintegrating rocks, which must always have been active upon the rocks protruded above the oceanic surface, and have produced continuous contiguous deposition of these rocks in another form at a lower level. This deposition, which is general over the floor of the ocean, does not necessarily at any period become visible as surface-rock unless it is elevated by plutonic forces above the level of the oceanic surface. The evidences of the fossil and other remains in the rocks, which remain permanent as it were by the accident of being projected above oceanic level, show in many cases that our visible surface-rocks may have been elevated locally by plutonic forces and have been degraded by meteoric forces many times before they produced any of the present finely disintegrated strata. The materials of the rocks appear after disintegration to be sorted out as it were, so as to depart entirely in structure from their original character, and therefore they must lose their former history.

The whole vestiges that remain of the long periods of the geological past represented by detached units, generally of slow deposition of mineral matter brought down by the action of rain, snow, wind, and tides, together with the large accumulation of remains of animal and vegetable life that have escaped after degradation, amount altogether to some 18 to 20 miles in thickness of deposition only, and these in separate detached units are all we possess from which to draw any inference whatever of the long period of past geological time.

330. The above statement, as regards the extent of past geological time in its entirety, does not preclude our recognition, if we please to accept it, of the evidences of certain conditions that ruled for certain long periods for which we may attribute active causes. Thus, as an instance, during the great Silurian period before mentioned, we find locally even miles in thickness of finely and slowly-deposited rocks in even stratification in a system which appears to have extended to a greater or less depth over the larger part of the Northern Hemisphere. In this we find here and there ripple-marked surfaces left of the ancient quiescent oceans, and of rain-marks on the smooth sandy beaches of the period. We are, therefore, upon these premises bound to conclude that during this time we have a period of mild quiescent atmospheric conditions and of the minimum effect of tidal action, possibly with heavy rainfall, for which causes may be fairly suggested. This subject will now be considered generally by taking the separate early periods defined by modern geologists and endeavouring to correlate them with the periodic conditions which have been proposed, particularly with regard to the action of the effective radiation of the sun's heat and of his contemporary volume under circumstances which have been already discussed.

331. *Archæan Period.*—After the metallic nucleus of the earth was formed (§ 235), we should have the possible approach of lighter nebular matter and its union with the surrounding

oxygen and halogens present, entailing complicated chemical processes, the principal factors of which would be the condensation and deposition of oxides and haloids upon the cooler polar areas of the earth, as before proposed. The great atmospheric pressure would cause also the heavy rains of highly heated water in circumpolar areas as before stated. The hot rains under the great pressure would dissolve the siliceous, aluminous, and calcareous rocks that were already precipitated from the condensed nebula near the poles and those also that were crowding outwards in pressure-folds under the influence of gravitation whilst floating upon the central dense liquid metallic base-matter. By the hot rains the protruding rocks as they were constantly dissolved at their surfaces would be brought down into the hollows.

332. Afterwards, as the atmosphere cooled, the semi-liquid rocks would be gradually deposited in a crystalline form in hollows and lake-basins. These depositions from aqueous solution probably produced the foliated crystalline rocks of the period most largely in gneiss, but partially also in mica and hornblende schists, chlorite slate, and crystalline limestone.

333. The land-areas formed of newly-deposited rocks at the close of the Archæan period would be left elevated by outward pressure of plutonic forces through polar pressures to great height at a distance around the poles. These rocks being constantly overflown with hot water left a system of permanent rocks, as before stated, locally thrown up at any point of minor resistance upon the first thin crust of the earth, or retained at the lower levels deposited from solution from its saturated mineral waters. The constant outflow of the lower heated viscid rocks moving into equilibrium to gravitational position, caused the upheavals from resistance to take place more particularly at the rounded borders of continental lands, from causes already pointed out. The general landsurface, therefore, was thereby covered with lake-basins formed

by surrounding mountains most elevated at the polar sides. These basins as the temperature lowered held the deposits of mineral matter brought down by the hot rains, particularly from the poleward slopes, where rainfall would be heaviest.

334. The entire system of what we now term the Archæan rocks is here assumed to be of the primitive rocks, which formed by themselves, and by after-degradation and ultimate deposition, the entire surface-rocks of the globe, as there could be no further deposition of mineral matter after the deposition of oxides and haloids from the nebulous atmosphere.

This system of rocks, including the deepest or lowest stratum, which still probably retains a white heat, could not have been less than 100 miles in thickness. The chilled surface-rocks of the Archæan system, left measurable at the present time in Canada, the Outer Hebrides, Bohemia and Bavaria measure about 40,000 feet in thickness.

335. *Archæan Time in relation to the Conditions of Animal Life.*—As the lower or early-deposited basic rocks not exposed to surface-radiation would remain for a long period in a hot viscous state, they would continue to flow slowly to a position of gravitation-equilibrium, distorting and crowding up all the new cooler incipiently-formed strata. It is during this period, which would have extended until the commencement of the clouding effects produced by the condensation of the nebular zone of Venus, that I propose to place the complete formation of our recognizable Archæan rocks. This period possibly lasted altogether from that following the liquid condensation of the metallic core of our globe for about 84 millions of years. This would complete nearly the entire first bright period, during which time the earth would remain in too highly heated a condition to maintain life upon its surface. Most probably during the latter part of this period, perhaps for 30 millions of years or longer, completing the 114 millions of the first period of our table (p. 225), the polar regions may have decreased in temperature by the

excess of polar radiation cooling the superincumbent vapour into rain-clouds of sufficiently low temperature upon precipitation to allow the commencement of organic life in polar regions. This new life, the cause of the beginning of which must rest in the Great Unknown, would be afterwards slowly distributed in succession further and further from the poles as the latitude-temperature fell by the continuous secular cooling at greater distance therefrom. The distribution under changes of circumstances would cause local variation of species outwards from the pole for adaptability to the surface-conditions; but as life-variation is a very slow process there would still be strong affinities in the newer animal life as it extended, with that which preceded its departure from its primitive polar home.

336. The Archæan rocks are here taken to be the earliest chilled superficial rocks, for the greater part projected to the surface by plutonic forces and acted upon by the ruling conditions present; but if we consider that the same forces are still evidently active to a certain extent in producing plutonic and volcanic phenomena, we can but take it that the system is more or less a continuous one lasting until the present time. The outward form of these rocks, particularly the mode of crystallization, must certainly vary in time from the differences of surface-conditions of the globe, in the heat maintained at the surface, the pressure of the atmosphere, the amount of aqueous vapour, and the mass of such elements projected to or beneath the surface at any period; but the materials must vary very little in chemical composition, coming, as they are here suggested to do, from the same universal source of exterior nebular deposition.

337. As we leave the Archæan period, we have new lights breaking upon us to guide us to the evidences of past time shown in the somewhat systematic evolution of organic life. But in this we have to contend with a constant erasure of evidences through the activity of the atmospheric phenomena

within any period, as before stated, which may leave us but faint indices of the changes through which the life and materials of surface-rocks may have passed before the rocks that remain to observation were formed. Thus in the 40,000 feet or so of old Cambrian and Silurian rocks to be presently considered, there is little evidence that we can derive from our experience of rock degradation and deposition that these rocks were formed directly from Archæan rocks. Indeed we must almost conclude from the average fine structure of the Silurians that they may have been churned up by atmospheric forces very many times during the Archæan period before they arrived at the state in which we find them in the Silurians. What we appear to possess indices of in rock-texture in many cases, is that there were within the Archæan period some special periods of deposition, separated possibly by longer periods, wherein rocks were deposited and disintegrated many times before they formed the strata we find open to observation which may have been formed by a special local deposition at a certain period. It is also probable that these periods of deposition were apart from any uniform conditions and depended greatly upon certain special conditions that were active upon the globe from the astronomical causes already discussed.

338. *Cambro-Silurian Period.*—Long before the condensation of the Venus-zone previously described, nebulous matter condensed interior to the earth's orbit would be floating in spiral bands towards this zone, sunward, obscuring a large part of the light and heat of the then large nebulous sun, so that the obscuration caused by the condensation of the Venus-zone when it was abandoned by the sun might not be very sudden; but as this zone cooled, the obscurity would constantly increase up to a certain point of its condensation, possibly even so far cooling the earth as to produce a temporary glacial period about the poles.

339. At the commencement of the formation of the Venus-zone, the vapour-laden atmosphere would exert a pressure

much greater than the present. The heavy temperate rains that fell at first over the polar areas only, would gradually extend over the present temperate regions, by which the surface-rocks would be worn off and washed down to the seashores, and pelagic life spreading originally from the poles would inhabit these temperate shores. The shores, by the excessive rainfall would be again covered with new deposits, brought down from the still elevated Archæan rocks, in which the mineral remains of organic life would be buried until the sea-bottom extended outward to a great area from the coast. These deposits would be again further diffused by oceanic currents. The sea would constantly but slowly rise by the amount of deposition in proportion to the inland denudation, and in a small degree by the loss of vapour in the atmosphere, and overflow its former boundaries.

340. The polar regions, by the effect of constant rainfall, would become degraded, as they would be also falling in temperature through the obscurity of the sun, until in process of time the cold would be sufficient at the most obscure period to cause life to become extinct through wide circumpolar regions. At the same time, under diminished heat from the sun, marine life, spreading outwards over the then temperate latitudes from the polar regions, would be supported by convection currents from the warmer submarine rocks not yet cooled to surface-temperature, so that life would extend over the entire temperate and tropical regions of the globe.

341. In the most obscure part of the period the initial heat of the globe, superficially cooled down, would maintain the deep waters in the tropical and temperate latitudes upon its surface at a fairly equable temperature favourable to marine life. Although under these conditions, if by ebb of tides or other circumstances such marine life was exposed in restricted areas in circumpolar regions without the circulation of warm currents, it would be destroyed by the low temperature

GEOLOGICAL PERIODS. 237

from the little heat given by the obscure sun, particularly in the winter solstice.

Towards the time of extreme obscurity of the Venus-zone, copious snowfalls would occur over the polar regions, which would press upon the chilled surface with sufficient weight to react upon the still semi-fluid siliceous rocks immediately underlying the surface-strata, so that these rocks might be pressed outwards in under-currents and underflow the surface-rocks, causing extensive local and mountainous elevations over many regions. These lower heated rocks would also overflow the earlier rocks in some districts by exudation due to reaction of polar pressures.

342. I place the duration of the Cambrians and Silurians, some time antecedent to and throughout the first dull period given in our table, p. 225, during which time the earth was sufficiently cooled down to permit the spread of marine life, at 96 millions of years according to our scale. This period to the field-geologist may appear relatively short to represent the great extent of deposition, which is locally in some districts 40,000 feet or more. We must, however, in this case consider the active state of the atmospheric conditions present, which would produce a nearly constant rainfall over a large part of the globe, and therefore that it was a period of constant deposition.

343. We have further the probability that the subaqueous equatorial regions of the earth continued sufficiently heated near the solid surface through all this period to produce a constant copious evaporation of the equatorial oceans, so that from this cause again the temperate regions would receive excessive rainfall, which would flow down to the warm oceans carrying mineral matter. The time, again, appears short for the evolution of the great quantity and variety of organic remains of pelagic life that we find in this period. But here, again, we have the probability of life having originated much earlier over the cooler polar regions in later Archæan times,

as before stated, § 153, and the possibility at this time, from the large disc of the dull nebulous sun subtending an angle, as proposed, of 67 degrees, that the small amount of heat derived from the sun was so distributed over temperate regions that it was particularly conducive to the rapid spread of marine life in the constantly subaqueously-warmed seas. In fact it was, as the fossil remains indicate, a perfectly equable quiescent temperate time, although under very constant rainfall—except only that the polar regions in the most obscure time may have been deeply covered with snow and ice, as before stated, to produce about these regions a temporary glacial period with contemporary elevation of surface-rocks elsewhere upon principles already discussed. During the publication of this work it has come to my notice that the equable condition of Silurian time, as being due to a large nebulous sun, has been suggested by M. C. Wolf in his able work ' Les Hypothèses Cosmogoniques,' p. 32.

344. The general evidences of the Cambro-Silurian time indicate that comparatively shallow seas prevailed or extensive flat shores which abounded with life. The frail shells indicate quiescent oceans free from storms and from much tidal action. These shores were constantly thickened by new deposition of fine sands, mud, or calcareous concretions, consequently to remain shallow they must have sunk or the ocean must have risen. Most probably both these conditions occurred. The recently highly-heated rocks would be slowly shrinking by loss of heat in superficial strata; and at the same time in this early age there would be a slow underflow of the still viscid lower heated rocks that were moving equator-ward to gravitation-equilibrium. At the same time the excess of rainfall would be lowering the surface-rocks, and in this degree raising the oceanic level as before stated, the whole condition causing great depths of shore deposits. The ripple-marks which give the evidences of quiescent times, independently of the even stratification, were possibly preserved to us by

occasional light clouds of dry dust blown over them from the higher lands during summer.

345. *Devonian Period.*—When the nebular zone of Venus had become broken and had commenced to condense to a nebulous planet of large diameter its disc would alternately obscure the sun and leave it open. At such a time the intensity of the heat and light of the sun would be intermittent in periods of the synodic inferior conjunctions of Venus. Such intermittent periods, by causing periodically rapid changes of temperature from mild to intense cold, would render the land-surface incapable of maintaining either vegetable or animal life, and be also destructive of littoral mollusk life. But in the open ocean, where such atmospheric changes due to temperature would create great disturbance, and thereby increase of the oceanic circulation, deep-sea mollusks and fishes of migratory habit could survive. It is at this period we have possibly the early progressive evolution of distinctly vertebrate fishes, which appear to be in a certain degree allied to the later forms of crustaceans and to the embryo forms of the future reptiles— possibly evolved partly, I would suggest, by extension of the caudal parts of these early animals, so that certain early crustaceans represent the head only of the later fishes. These fishes, which came to perfect development in the period represented by the Old Red Sandstone in Great Britain, were evidently forms adapted to the required conditions of migratory habit. They possessed in most cases powerful fins and tails, the fins in some cases being possibly adapted to rapid surface swimming, such fishes of the entire family of the Asterolepidæ, as *Coccosteus decipiens*, *Pterichthys oblongus*, *Cephalaspis Lyelli*. Other fishes were evidently adapted to deep-sea swimming, as *Holoptychius nobilissimus*, *Osteolepis major*, and many other forms, all of which were equally adapted to migratory life.

346. After the intermittent obscure period there would be

a period in which the condensing globe of Venus would become neutral in its heating effects, being neither obstructive nor auxiliary to the sun's heat. At this period we have all the conditions conducive to the spread of vegetable life over the land-surface.

347. From the period of greatest obscurity by the condensation of the nebular globe of Venus to that of neutrality of this planet's effects upon the sun's radiation, would be the time of greatest deposition of snow upon the cooler portions of the earth during the winter solstice. Through this period the earth's crust, being much thinner than at present, would be more sensitive to upheaval and intrusion of plutonic matter beneath the surface from pressure of the snow at the poles bearing down the polar surface rocks. There would therefore be evidence of plutonic action, taken in its broadest sense, almost synodically during this period, elevating immense districts of land-areas in steps.

When Venus had condensed to an incandescent state, plutonic and volcanic action would cease and the entire snow at the poles would be melted, so that there would be a return to purely aqueous conditions with elevation of sea-level. The entire period represented by the Devonian in Great Britain I assume to have lasted about 60 millions of years, extending beyond the third period of our table.

348. *Carboniferous Period.*—This would commence in the intermittent bright fourth period of our table. The temperature of the air would be raised at the period of internal conjunction of Venus and gradually fall, so that there would be intermittent periods of bright sunshine followed by heavy rains. This would produce thin intermittent stratification of rocks upon the shores and in inland lakes. These lakes would overflow at the rainy period and cut incipient river-channels, leaving thereby extensive marsh-lands, which would represent in series extending coastwards the rivers of the time. When the sun became open in the fifth period these marsh-lands,

GEOLOGICAL PERIODS. 241

owing to the uniform temperature and general quiescence of atmospheric conditions, would become crowded with vegetable life, and the general conditions would also be conducive to the evolution of amphibian life. The Carboniferous period would extend throughout the entire fifth open period of about 306 millions of years until the commencement of the clouding of the sun for the commencement of the formation of the Mercury-zone.

349. The long period suggested above may be thought to exceed that demanded for Carboniferous time by the evidences presented by geology. The strata represented appear to be only from about 2000 to 13,000 feet where deposits are left. We must observe that the deposits are generally very fine, and indicate slow intermittent formation frequently of fine windborne sand without animal life, or layers of fine mud that was disintegrated by gentle rains from the rocks into the shallow surface pools. These were possibly filled with delicate cellular animal and vegetable life, which has left no remains. Such light strata were again easily eroded after deposition under plutonic elevation, and have no doubt for the greater part entirely disappeared, being incorporated into later formations.

350. *Permian Period.*—When the sun was obscured by the incipient condensation of the Mercury-zone, the conditions given for the first dull period when the Venus-zone was formed, which commenced the Cambro-Silurian times, would be nearly repeated, but now the terrestrial conditions under which it was superimposed would be entirely different. The earth's crust would have greatly thickened by cooling, so that the tropical ocean-bottom would no longer heat the superimposed waters in a sensible degree to cause great evaporation over these regions, as in Silurian times, and heavy warm rains over polar regions therefrom. The general surface-strata would also be cooled down, so that the conditions of cooling which in the case of the Silurians induced

R

the spread of pelagic life would in this latter case cause its destruction. The general cooling of the earth at this period by the obscuration of the sun's heat therefore probably caused a steady drain of vapours from the surface of the ocean, seas, and lakes, which was deposited as snow upon the higher land-surface. The lakes on evaporation left saline deposits in their beds, combined with fine dust or sand blown from the higher rocks disintegrated by frost. The accumulation of snow over all the present circumpolar temperate lands caused, by its pressure upon the still semi-fluid lower rocks, the general elevation of lands of the earlier periods more distant from the poles, floating these up bodily in some districts and by local disturbance with unconformity in others.

351. It is during the Permian period as it is represented in Great Britain that we find the dying out of palæozoic life. In the early most obscure period possibly the conditions were such that life became extinct within the entire present temperate regions of the globe, except in the deep waters of the ocean. Upon the earth's surface within the tropics alone, where the dim sun could effectively diffuse its direct though obscure rays, prevalence of life was possible.

352. In the continuity of life after the dull period we have directly the reverse conditions to those that ruled after the earlier dull Silurian period. We have life spreading from the tropics instead of from the poles. The migratory species that could reach the temperate shores were also generally much more highly organized and locomotive, and adapted to bring in the important factors of animal existence in the new era of the Mesozoic period.

This dull Permian period would break into the period of the separation of the Mercury-zone, by which an intermittent period would be caused, again producing rapid changes of thin stratification of rocks, not at first conducive to the existence of organic life till the condensation of Mercury ceased to obscure the sun. What we may possibly define as the Permian period lasted nearly through the sixth dull period, which, according to our table, p. 225, would be about 60 millions of years.

353. *Triassic and Rhætic Periods.*—This division would complete the second dull to bright intermittent period, when Mercury was formed as a nebulous planet obscuring the sun when in inferior conjunction. The rapid changes produced in temperature would cause rapid deposition of rocks and greatly restrict the conditions of organic life; but gradually the volume of Mercury would become less before entering into the state of an incandescent planet and more calm conditions would prevail. The Triassic period would last about 47 millions of years, completing the open to dull period.

354. *Jurassic Period.*—The eighth of our table or the second auxiliary period. We again enter into a time particularly adapted to the rapid evolution and spread of pelagic organic life. The sea-shores by the intermittent rainy conditions due to temperature changes, yet always warm, became covered with fine mud from the land-areas, bringing down with it abundance of support for a suitable class of littoral life, carrying at the same time into the deeper oceans carbonate of lime in solution available for all forms of life depending upon it. This period would extend throughout the eighth division of our table, for about 37 millions of years.

355. *Cretaceous and Eocene Periods onwards to the present Time.*—These may have been partly contemporary if we imagine that the Chalk was a formation more distant from the ancient shores, which was continuously over-deposited in some areas, while elsewhere it was covered by argillaceous rocks, as the land-surface was degraded. This is taken to represent the ninth period of our table. This period, under constant diminution of the sun's disc, would be generally equable, producing only gentle deposition of rocks locally in drainage-basins or in the distant ocean-bottom. It would

otherwise be only subject to the set of changes that were due to the condition of the sun at the critical temperatures of its former nebulous surroundings, which may have produced occasionally more or less obscure periods, to which we may add the changes due to eccentricity of orbit and variation of axis, which have produced great changes in climate. This long period drifts us into the present time, when the sun's disc has become very small relatively to what it was in the past and of intense light and heat.

The period set aside by our table to include the Cretaceous and Tertiary periods is 532 millions of years. This may appear long for the number of superficial changes that are evident upon the surface from the amount of mineral matter deposited. In such a uniform period the variations were probably all local and for the most part intermittent through the periodical astronomical changes, so that land was alternately deposited and degraded many times without producing any great entire depth of strata*.

356. With regard to life, our only true index of time in this period, we may conclude that where there is a general constancy of like conditions there is little reason for change for adaptibility to the circumstances present, particularly if the organism is elevated to a condition of migratory instinct for accommodation to the seasons. Therefore, seeing that the changes have been great, the evolution period must have been immense to have produced the variation in forms which we know to have occurred in this period, more particularly in that of the elevation of the scale of the higher mammals. The changes within the tertiary period present wonderful variations in the structure of the mammalia to give the number of species we at present possess without consideration of the number extinct. There is one mark, however, of continuous progress throughout all this period—the brain constantly

---

* Appendix C.

grows larger in relation to the bulk of the animal. This is possibly a very slow continuous progressive feature, which may give us some idea of the enormous extent of time embraced within this period.

357. We may generally postulate from geological evidence that in the long Tertiary period we have had present the conditions proposed by Lyell, in his 'Principles of Geology,' for the entire system of deposition of surface rocks, wherein every deposit is assumed to depend upon local circumstances, and the rocks of the past to be simply disintegrated and redeposited without any important change in the cosmic state.

358. In the above discussion of periods the effects of changes of eccentricity of orbit and variations of the obliquity of ecliptic have only just been mentioned. These variations, there is no doubt, produced marked effects and caused changes of climate locally which have produced minor divisions in stratified rocks. This subject has been ably discussed by others, and probably in some cases its importance has been much exaggerated *.

359. There are also many conditions that have materially affected the contemporary stratification, which depends greatly upon the sea-level of the period, some of these have been considered, but they still leave generally a wide field for future investigations. The most important is the study of the equilibrium of the earth's mass as a gravitation system in rotation under the superficial changes of surface rocks that are evident in periods of the past; thus, for instance, the greater or less elevation of ice at the South Pole would disturb the gravitation centre. A past reaction of ice at the North Pole, which probably, upon conditions proposed above, was the direct cause of elevation of the entire plateau of

---

* Lyell's 'Principles of Geology,' vol. i. p. 272 ; Author's paper, British Association Reports, 1884, p. 723.

Central Asia. This would again react on the earth's equilibrium. Both of these systems would tend to lower the sea-level of Great Britain; but the subject is too large to be even sketched in the present treatise.

360. *Glacial Period.*—As regards this period, which necessarily comes within the ninth division, this, as a consequence of the natural diminution of the sun's volume, has been discussed in Chapter XIII. My early reflections upon this subject, when strongly imbued with the Huttonian principles developed by Lyell, led me to think that there were probably indices of sufficient change under local conditions with slight variations of elevation of land-areas to account for this epoch in Europe. These conclusions were mainly based upon my idea that the movements of aerial and oceanic currents are caused by the expansion of the atmosphere, and in less degree of the oceanic surface, by the heat of the sun in its apparent diurnal motion through the tropics driving before it an expansion-wave of air and water *. This as a systematic motion, according to my theory, would produce whirls or cyclones lateral to the tropics, the position of which would depend upon the resistance of the coast-lines. Under these conditions the resistances that locate the North-Atlantic whirl are the coasts of North America. The peculiar conformation of this coast at the present time deflects the North-Atlantic whirl into a bi-whirl, of which the northern part—that is, the Gulf Stream—crosses the Atlantic, passes along the coast of Norway, enters the Arctic Circle, and is deflected back to its origin after passing along the coast of Greenland, which it leaves glaciated by the cold Arctic current. This direction of rotation is quite abnormal to that of any other lateral tropical whirl. It is clear, as stated in my paper read before the British Association in 1885 †, that if

---

* 'Fluids,' p. 626.
† Brit. Assoc. Reports, 1885, p. 1020.

the lower central lands of North America were only sunk some 300 feet from the mouth of the Mississippi through to Hudson's Bay, a result readily produced by cosmic causes already discussed, this whirl would then take its normal course, as other lateral whirls in the South Atlantic, Pacific, and Indian Oceans. Under these conditions Northern Greenland would be placed in the tropical current and enjoy a very temperate climate, and the returning Arctic oceanic and aerial current would bathe the coasts of Western Europe, leaving icebergs on the coast with inland glaciation, as at present in Greenland. This would include also the glaciation of the northern part of Great Britain.

361. To account for the glaciation of Northern America upon like principles to the above, we should require the mean temperature of the Northern Pacific Ocean to be sufficiently high for its whirl deflection to keep the Behring Sea open, so that the Northern Pacific whirl could continue its direct motion in open water north of Alaska, bringing the return Arctic aerial currents into the valley of the Mackenzie River and through the Great Bear and Great Slave Lakes into the valley of the Missouri, laden with sufficient moisture to produce a contemporary glacial period in the northern parts of the South-western States and distribution by currents through the lateral valleys.

362. The considerations which have made me somewhat modify this idea, without change of the principles suggested so far as they are active, were due to a more attentive study of American geology. The glaciation of Arctic North America appears to have been greater than these principles would entail. This to my mind is seen most particularly, according to my theory, in the evidence of the great outflow of basalt in the region of the Snake River, Idaho, within tertiary times. To produce this great outflow of lower heated rocks upon principles herein discussed, there must have been very great elevation of ice, most probably about the North Pole.

I assume that a pressure system of ice at the poles would react throughout the entire lower viscous rocks; but as these rocks are assumed to be supported by flotation upon the denser metallic nucleus, the reaction by protrusion to restore gravitation-equilibrium at a distance from either pole would be much more frictional than that from a nearer pole. The mass of basalt protruded in Idaho possibly equals the entire mass of land above the oceanic surface in Great Britain *. There was also contemporary elevation of the great plateau of Central Asia. There may therefore have been an ice-cap in the north, somewhat equivalent to that of the Antarctic Circle. The cold necessary to produce such an ice-cap we can scarcely imagine to have existed at the present mean temperature of the globe. It is therefore more consistent to assume that the effective radiation of the sun was diminished for a period, probably, as I have proposed, by clouding in the condensation of nebular matter at its critical temperature, p. 74. This cannot, however, destroy the evidence of glaciation, being at any time local in intensity, and the marine shells in the glacial clays indicate an open ocean poleward †. There is said to be no evidence of glaciation in the great plains of Siberia, and from my own observation there has been none in the west of Norway; for instance, upon the granitic and gneiss rocks of the Lofoten Isles which retain the sharp pointed outlines of ancient rocks that have been subject to weathering only throughout long geological periods ‡.

For great elevation of circumpolar land or massive outflow of basalt at any period, the extreme cold of the previous period may have produced great rigidity in the ice-system and contiguous rocks. The reaction of such a system would cause the more distant parts of the earth's crust to give way

---

* Geikie, 'Geology,' p. 257.
† 'Acadian Geology,' Sir J. W. Dawson, p. 65.
‡ Author's paper, Geol. Soc. Proc., Feb. 23, 1887.

paroxysmally and after fracture to continue the effects of the reaction until the polar pressure reached nearly its point of equilibrium. In this manner reactions of ice-pressures would be periodical upon the surface system of rocks.

363. *Future Period.*—By the continuity of the conditions which now rule, the sun will not probably become a much brighter incandescent body than it is, and it will decrease in volume and ultimately in incandescence. The ice-caps which cover the Antarctic and probably the Arctic pole must therefore grow more extensive with the decrease of solar heat ; and although increase of weight of ice may cause deflection of the crust and distribution of pressure upon the interior highly heated rocks, which pressures may react in producing plutonic and volcanic phenomena, still, with each such displacement the crust will become thicker, more rigid, and more resistant, and thereby volcanic and plutonic action to overcome the resistance will be more paroxysmal. The oceans also, by the condensation of evaporated water and its deposition in the form of ice at the poles, will become of less depth, so that land-areas will increase in aerial surface.

364. The periodic process would therefore appear to be a general decrease of temperature accompanied by apparent elevation and spread of continental lands and decrease of oceanic area, so that the temperate inhabitable globe would increase over tropical areas for a long period in greater ratio, possibly, than the circumpolar areas and would cease to be sufficiently temperate for the existence of organic life.

At a later period the evaporation from the tropical oceans would be less, and deposition in circumpolar areas less, and the earth's crust more rigid, so that the sea would diminish less by this cause; but at the same time the mountainous lands surrounding the great oceans would have the snow-line lowered, so that a part of the evaporation from the oceans would fall upon the adjacent lands instead of drifting poleward, and the lower shelving shores of the much reduced tropical oceans would become habitable lands.

At a still later period the tropical ocean-beds would be drained by evaporation in clouds drifting over to the shores, which for the most part would never return to them in the frozen river-streams. The earth would then possess three or four oceanic areas adapted to life only—the lower beds of the Pacific, Atlantic, and Indian Oceans.

Later when the sun presented only a dull red disc, appearing to move daily across the starry vault, the lands representing the deep ocean-beds would be frozen and life gradually become extinct.

365. The entire cloud-drainage of the great oceans and the snow-clad mountainous lands surrounding them would release the pressure upon the deep ocean-beds in proportion to the increase of the weight of snow upon the mountains. The earth would probably still be sufficiently yielding in its interior to admit of a certain amount of reaction by distribution of surface-pressures, by which the ocean-beds would be elevated to restore partial gravitation-equilibrium. This effect would probably be produced by distribution of small volcanoes over the former lower oceanic surface, and, as the surface would be shrinking slightly by loss of temperature, there would occur also paroxysmal upheavals in localities where, through the tension and plutonic pressure below the surface, resistance would be overcome. This would leave earthquake-fissures as a permanent surface-feature.

366. It is possible that in one of the warmer tropical valleys running east to west formed from one of the above-described fissures in the deepest bed of the Pacific, and near to some residual thermal springs, the last individual, of the latest evolved form of humanity, may die of hunger and close for ever the records of science attained upon our globe.

Whether extinction of life will occur within the short period of 15 millions of years, as suggested by the theory of

Helmholtz for the sun to decrease to the density of the earth, can scarcely be suggested. Whether we know the sun's specific heat or the law of dispensation of solar heat into space is doubtful ; but the probability is that world-life will be longer than this, if we can accept the conditions which I have reserved for discussion in Appendix A—as my theory on this point may be thought not to be a necessary part of our subject.

APPENDIX A.

I MAY offer as a pure hypothesis that the energy of a light- and heat-giving system may not be so rapidly dissipated as we know it to be by experiment unless it meet in radiation with another material body as a recipient. In fact, that for the rapid diffusion of light and heat through cosmic space there must be one or more motive couples, just in the same way as this is necessary for the action of gravitation-energy, only that light and heat bear reference to surface only, not to mass. Upon this hypothesis light and heat, as possibly gravitation, may be considered in certain cases as phenomena of *induction*, and in action, in a certain degree as regards the sun, resemble the discharge from one excited conductor to another through an insulated medium. The intensity of propagation of forces from the sun's globular mass being assumed equal to that of the discharge of electricity from a point into a space of direct insulation—that is, insulation from general diffusion,—so that the induction to another body if present falls in direct line only, with the loss only of a limited amount by free radiation. In this case the form of force is nevertheless that of light or heat, not of electricity, from which it may vary in any motive degree. A theory upon these lines, to which I have devoted some years of study, but cannot extend here, except to state the principle, would account for there being snow-caps about the poles of Mars, whereas, from his distance being taken inversely as the square in comparison with that of the earth, the amount of sun-heat capable of reaching the surface of this planet would leave it wholly frozen. The intensity of the light of Jupiter and Saturn also far exceeds that due to the reflection of the sun's light as deduced from calculation of uniform radiation only. Upon these principles, if they hold, it may be suggested that we should materially conserve solar energy, for however closely light induction couples may fill up the radiants about the sun, the open interspaces where there would be less loss of energy must be of immensely greater area—that is, as the space to the mass. Further, the sun would receive as much light-energy from another star as it would impart to this star, so that the radiant force dissipated would be principally upon the planets—that is, upon bodies cooler than itself. Such a principle of dispensation of heat and light would prolong the sun's future life to many times the period estimated by Helmholtz.

To separately define the forces active upon a planet in relation to the sun in factors of induction according to this hypothesis, I would suggest :—

Gravitation.—*Mass Induction,* producing a tendency to draw two bodies together whose equilibrium is only satisfied when the attraction produces a unit globular mass.

Heat.—*Conductive Induction,* by which a certain depth of mass is affected in diminishing ratio in intensity by some geometrical power. This force is probably, if taken *per se,* repulsive.

Light.—*Surface Induction,* affecting the surface molecules only to a limited depth, except in special or dioptric bodies, through which it passes to other bodies by a system of radial conduction to the inductive body beyond. This force is also *per se* probably repulsive.

As I may never publish my researches on this subject, I may say that my idea of light is that its induction renders all bodies under its influence luminous to a certain extent. That this self-luminosity induces a like secondary luminosity in another body or a similar effect upon the retina of the eye or a sensitive film in a camera. That this luminous action is probably caused by rotating the surface molecules so as to cause them to present the surface—which affinity in the dark draws inwards—to an outward position. A complete revolution being necessary for white, and a partial revolution for colour. But this last stated idea is immaterial, the self-luminosity is material. To mention one of my first experiments. I made in 1873 a drawer 6 by 6 inches, of one inch in depth, very carefully fitted in a velvet-lined frame to exclude light. The inside of the drawer could be exposed to sunshine while it was closed and drawn into a dark room when required. I attached two spiral springs to the front of the drawer and a catch to keep it closed when out in the sunshine, so that when the catch was released the drawer came instantly into the dark room. I tried many experiments; the first was that of writing my name boldly in Indian ink upon a piece of white paper. After this had been a minute or less in the sunshine, and was then drawn into the dark room, I found that I could read it easily for a short time in the dark ; therefore, I conclude it retained a *part of its luminosity*, or at least, if it was luminous in the dark, it must have been also luminous in the sunlight, so that its perception to the eye could not have been entirely from reflected sunlight as generally assumed. It appears to me, therefore, that we are bound to admit induced luminosity as a factor of visibility. These effects, as phenomena ascribed to phosphorescence, are well known, and have been investigated most ably by Becquerel ; but my idea of them is that they are not, as assumed, simply phenomena of phosphorescence, but of induced luminosity, and that they are universal for all light-giving bodies, that is, for all visible bodies or such as are not dioptric, or for black bodies, if any exist, or so far as they exist. In this hypothesis it is not necessary to suppose that a body may retain its induced luminosity for an instant in the dark.

Heat conductors, that is metallic bodies, possess no such power of retention, the surface molecule being assumed in this case to be sensitive instantly to light and heat influences. This does not affect the laws that govern the action of light, such as the reflective properties under the condition that a body may receive luminous induction in one direction and dispense it at coincident equal angles, or the refractive properties of dioptric bodies, by which the inducing rays are bent, only that in this last case the inductive body is behind the dioptric, which acts only as a conductor thereto in the same manner as a metal wire does to electricity, but following its own laws. The direction of the light force of induction is otherwise always in direct line. My experiments upon these hypotheses were made in 1872-4, and I have discussed some of them with my friends. Some of these ideas appear to have occurred to a correspondent of the 'English Mechanic,' T. W. B., last year, 1894, and, so far as that publication goes, I acknowledge the priority if it is of any value. In the multiplicity of thoughts, by reason of our similarity of organism, some of our ideas must be like those of other individuals.

APPENDIX B.

THE generally accepted theory of land-formation is that which was proposed or maintained by the late Robert Mallet in a paper upon " The amount of Energy developed by the Secular Cooling of the Earth," contained in two papers, over 100 pages, in the Phil. Trans. 1874-5. According to these papers the amount of heat lost from the initial temperature of the earth will represent the force of its contraction. The amount of this energy is presumed to be made evident in compression of the superficial strata causing the elevation, inclination, and crumpling of the strata and the entire volcanic phenomena. The data, upon which the arguments of these papers rest, are assumed to be taken from calculations of Elie de Beaumont, Forbes, and Lord Kelvin, who estimate the heat lost by the earth to be equal to the melting of a plate of ice, respectively of 0·0065, 0·007, and 0·0085 millimetres annually. From these data it is stated that from 575 to 777 cubic miles of ice melted annually would represent the loss of heat. By going over the calculations in this paper, I was able to point out a considerable error in it, sufficient to upset the whole contraction theory upon the lines laid down by Mr. Mallet. After writing to Sir George Stokes, then Secretary to the Royal Society, who clearly saw the accidental error, I read a paper upon it before the Geological Society * in June 1884, showing that the contraction from the data given was only about one cubic mile annually, that is,

* Quart. Journ. Geol. Soc. vol. xl. (Proc.) p. 67.

from ·7937 to 1·0387 mile. The principal authority for the data given was Lord Kelvin; and as I could not find any reference to the subject in his papers, Sir George Stokes kindly wrote to Lord Kelvin for me about this, and found that the assertion was altogether a mistake. Lord Kelvin never made such a calculation, therefore this theory, supported upon the evidence of compression of surface-strata, is generally without foundation in fact. I think, moreover, that the contraction theory is quite opposed to observation of actual rocks, the joints of which are generally open below the surface, and show the effects of pressure from beneath producing a *tensile strain* upon the surface to form the open joints. These joints are often filled with basalt from intrusion of the underlying magma, showing more directly evidence of the outward pressure of heated liquid plutonic matter.

APPENDIX C.

IF the general theory of this work is accepted at some future time, a more experienced practical geologist than myself may shift the divisions in the rock-series that I have adopted to make them more exactly agree with the periodic conditions proposed. To do this perfectly would require refined geological observation, as the astronomical changes herein defined could not have been generally abrupt, so as to produce very distinct divisions in stratifying rocks. Further, there must be superimposed upon the greater astronomical changes herein suggested, the minor influences of variation of eccentricity of orbit and change of obliquity of axis, which would produce variation of deposition although possibly not to the extent proposed by Croll and Lyell. Some objection, for instance, may be made to my grouping the cretaceous with the tertiary in one long period, wherein the chalk formation is at least very distinct, and locally no doubt, if taken in vertical series, the more ancient. In this case we may consider the chalk to be a deep oceanic formation that is still in progress, a theory generally accepted since the 'Challenger' Expedition. I think a system of contemporary stratification of the various kinds of sediment distinguished by special chemical elements must have been general throughout all time, as we have only one set of such elements largely to deal with upon the surface of the globe, however much they may have been churned up or sorted out by local prevailing conditions. Upon this suggestion we could at no period have had one general system of deposition prevailing either of silica, alumina, or calcic-carbonate, in other than local areas. The

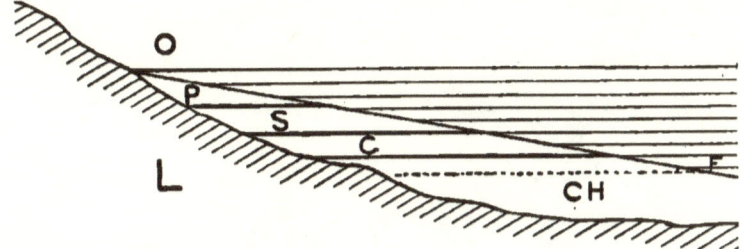

general scheme of deposition in quiescent times and undisturbed by oceanic currents may be shown diagrammatically by the figure above, which may represent, say 200 miles of, deposition from a coast of the ancient rock-surface of a certain period :—O, the oceanic surface; a line from the ancient rock to a point F the surface of the newly-formed rocks, where a band of flints occur in the chalk from organic deposition at a certain distance from the coast. Then of the produce of the disintegrated rocks, the coarser materials would rest at B; the broken masses of silica or sand at S; the lighter mud or clay at C; the perfectly soluble carbonate of lime and silica at CH, where it would be generally absorbed by organic life. This system would in all cases form *sets of rocks* and go on continuously over areas of surface-drainage carrying the disintegrated rocks if undisturbed by oceanic currents or tidal action, and could in the past only be arrested by such great astronomical changes as herein proposed. These greater changes cannot occur again, so that the present period may be geologically indefinitely extended for the time the ocean retains its liquidity.

www.ingramcontent.com/pod-product-compliance
Lightning Source LLC
Chambersburg PA
CBHW031933230426
43672CB00010B/1909